T0210547

SpringerBriefs in Mathematics

SpringerBriefs in Mathematics showcases expositions in all areas of mathematics and applied mathematics. Manuscripts presenting new results or a single new result in a classical field, new field, or an emerging topic, applications, or bridges between new results and already published works, are encouraged. The series is intended for mathematicians and applied mathematicians.

More information about this series at http://www.springer.com/series/10030

Ping Zhang

A Kaleidoscopic View
of Graph Colorings

 Springer

Ping Zhang
Department of Mathematics
Western Michigan University
Kalamazoo, MI, USA

ISSN 2191-8198   ISSN 2191-8201   (electronic)
SpringerBriefs in Mathematics
ISBN 978-3-319-30516-5   ISBN 978-3-319-30518-9   (eBook)
DOI 10.1007/978-3-319-30518-9

Library of Congress Control Number: 2016934706

Mathematics Subject Classification (2010): 05C15, 05C70, 05C78, 05C90

Printed on acid-free paper

This Springer imprint is published by Springer Nature
The registered company is Springer International Publishing AG Switzerland

Preface

It is the origin of the Four Color Problem by Francis Guthrie in 1852 that led to coloring maps and then to coloring planar graphs—not only coloring its regions but coloring its vertices and edges as well. In 1880, when Peter Guthrie Tait attempted to solve the Four Color Problem, it was known that the Four Color Problem could be solved for all planar graphs if it could be solved for all 3-regular bridgeless planar graphs. Tait was successful in showing that the Four Color Problem could be solved in the affirmative if it could be shown that the edges of every 3-regular bridgeless planar graph could be colored with three colors in such a way that no two adjacent edges are colored the same. He did this by showing that such an edge coloring of these planar graphs led to a coloring of their regions with four or few colors so that no two adjacent regions are colored the same, and conversely. While Tait's approach never led to a solution of the Four Color Problem, his idea of how one coloring of a graph can lead to another coloring of value has opened up a large variety of coloring problems. The major goal of this book is to describe the kaleidoscopic nature of various colorings that have been studied in graphs.

Over the years, there have been many graph colorings that have led to other graph colorings of interest in a variety of ways. In the author's book *Color-Induced Graph Colorings*, various edge colorings were described that result from vertex colorings of interest. In this book, this topic is continued. While we will be describing many ways that edge or vertex colorings have given rise to other colorings and discussing some of the major results, problems and conjectures that have resulted in this area of study, it is not our goal to give a detailed survey of these subjects. Indeed, it is our intention to provide an organized summary of several recent coloring concepts and topics that belong to this area of study, with the hope that this may suggest new avenues of research topics.

In Chap. 1, the background for basic colorings concepts is reviewed. There is also a review of a number of other concepts and results that will be encountered throughout the book. In Chaps. 2 and 3, edge colorings of graphs are discussed that lead to vertex colorings defined in terms of sets and multisets of the colors of the edges. This leads to colorings called binomial colorings, kaleidoscopic colorings, and majestic colorings.

In Chaps. 4 and 5, we discuss vertex colorings that induce a variety of edge colorings, which are related to the well-known graceful labelings and harmonious labelings.

In Chap. 6, region colorings of planar graphs are discussed, where regions sharing a common boundary edge are required to be colored differently. Several region colorings are described that not only distinguish every pair of adjacent regions but which potentially require the use of fewer colors than a standard region coloring. This, in turn, leads to vertex colorings of graphs in general, discussed in Chaps. 7– 10, which are, respectively, defined in terms of sets, multisets, distances, and sums of colors.

In Chap. 11, two combinatorial problems are described, leading to two graph coloring problems, which are also discussed in this chapter. In Chap. 12, two Banquet Seating Problems are described, each of which can be modeled by a graph and suggests two types of colorings of the graph. This gives rise to two vertex colorings of graphs in general, which are the topics discussed in Chaps. 13 and 14.

Kalamazoo, MI, USA   Ping Zhang
21 December 2015

Acknowledgements

With pleasure, the author thanks Gary Chartrand for the advice and information he kindly supplied on many topics described in this book. In addition, the author thanks the reviewer for the valuable input and suggestions on the first draft of this manuscript. Finally, the author is so grateful to Razia Amzad, SpringerBriefs editor, for her kindness and encouragement in writing this book. It is because of all of you that an improved book resulted.

Contents

List of Figures

Chapter 1
Introduction

One of the major areas within graph theory is that of colorings, namely region colorings of graphs embedded on surfaces, vertex colorings and edge colorings. Of all these colorings, the most studied and most popular graph colorings are the vertex colorings. These colorings came about through coloring the regions of planar graphs, that is, through coloring the regions of maps. In this chapter, we review some fundamental concepts and results on vertex and edge colorings that will be encountered as we proceed. In addition, we review some facts concerning degrees of vertices in graphs, outdegrees and indegrees of vertices in digraphs as well as Eulerian graphs and digraphs. Finally, a fundamental fact from discrete mathematics is mentioned that will be encountered often. We refer to the book [15] for graph theory notation and terminology not described here.

1.1 Graph Colorings

It was 1852 when the young British scholar Francis Guthrie brought up the question of whether the regions of every map could be colored with four or fewer colors so that every two regions having a boundary line in common are colored differently. This *Four Color Problem* has a lengthy and colorful history (see [75]). While a solution to this problem would not be found until 1976, there were many attempts to solve it by many individuals during its 124-year history. A famous incorrect (although interesting) argument was made by the British lawyer and mathematician Alfred Bray Kempe in 1879. Kempe observed that coloring the regions of maps so that "adjacent regions" are colored differently was the same problem as coloring the points of certain diagrams so that two points joined by a line are colored differently (coloring the vertices of a planar graph so that adjacent vertices are colored differently).

© The Author 2016

P. Zhang, *A Kaleidoscopic View of Graph Colorings*, SpringerBriefs in Mathematics, DOI 10.1007/978-3-319-30518-9_1

Thereafter, the study of vertex colorings of graphs in general in which adjacent vertices are colored differently (proper vertex colorings) has become a major area of study in graph theory. Over the years, there have been numerous changes in properties required of vertex colorings and in ways that certain graph colorings have resulted from other graph colorings. In [76], various edge colorings were described that result from vertex colorings of interest. In this book, this topic is continued. However, the major goal of this book is to describe the kaleidoscopic nature of vertex colorings that have been studied in graphs.

1.2   Proper Vertex Colorings

A *proper vertex coloring* of a graph G is a function $c : V(G) \rightarrow S$, where in our case, $S = [k] = \{1, 2, \ldots, k\}$ or $S = \mathbb{Z}_k$ for some integer $k \geq 2$ such that $c(u) \neq c(v)$ for every pair u, v of adjacent vertices of G. Since $|S| = k$, the coloring c is called a *k-vertex coloring* (or, more often, simply a *k-coloring*) of G. The minimum positive integer k for which G has a k-vertex coloring is called the *chromatic number* of G, denoted by $\chi(G)$. Suppose that c is a k-coloring of a graph G, where each color is one of the integers $1, 2, \ldots, k$ say. If V_i ($1 \leq i \leq k$) is the set of vertices in G colored i (where one or more of these sets may be empty), then each nonempty set V_i is called a *color class* and the nonempty elements of $\{V_1, V_2, \ldots, V_k\}$ produce a partition of $V(G)$. Because no two adjacent vertices of G are assigned the same color by c, each nonempty color class V_i ($1 \leq i \leq k$) is an independent set of vertices of G.

For graphs of order $n \geq 3$, it is immediate which graphs of order n have chromatic number 1, n or 2. A graph is *empty* if it has no edges; thus, a *nonempty graph* has one or more edges.

Observation 1.2.1. *If G is a graph of order $n \geq 3$, then*

(a) $\chi(G) = 1$ *if and only if G is empty.*
(b) $\chi(G) = n$ *if and only if $G = K_n$.*
(c) $\chi(G) = 2$ *if and only if G is a nonempty bipartite graph.*

An immediate consequence of Observation 1.2.1(c) is that $\chi(G) \geq 3$ if and only if G contains an odd cycle.

The *union* $G = G_1 + G_2$ of G_1 and G_2 has vertex set $V(G) = V(G_1) \cup V(G_2)$ and edge set $E(G) = E(G_1) \cup E(G_2)$. The union $G + G$ of two disjoint copies of G is denoted by $2G$. Indeed, if a graph G consists of k (≥ 2) disjoint copies of a graph H, then we write $G = kH$. The *join* $G = G_1 \vee G_2$ of G_1 and G_2 has vertex set $V(G) = V(G_1) \cup V(G_2)$ and edge set $E(G) = E(G_1) \cup E(G_2) \cup \{uv : u \in V(G_1), v \in V(G_2)\}$. The *Cartesian product* G of two graphs G_1 and G_2, commonly denoted by $G_1 \,\square\, G_2$ or $G_1 \times G_2$, has vertex set $V(G) = V(G_1) \times V(G_2)$, where two distinct vertices (u, v) and (x, y) of $G_1 \,\square\, G_2$ are adjacent if either (1) $u = x$ and $vy \in E(G_2)$ or (2) $v = y$ and $ux \in E(G_1)$. The definitions of the union, join or Cartesian product of two graphs can be extended to the union and join of any finite number of graphs. If a graph G is the union or the join of k graphs $G_1, G_2, \ldots, G_k$ for some integer $k \geq 2$, then the chromatic number of G can be expressed in terms of the chromatic numbers of these k graphs.

Theorem 1.2.2. *Let $G_1, G_2, \ldots, G_k$ be k graphs where $k \geq 2$.*

(i) If $G = G_1 + G_2 + \cdots + G_k$, then $\chi(G) = \max\{\chi(G_i) : 1 \leq i \leq k\}$.
(ii) If $G = G_1 \vee G_2 \vee \cdots \vee G_k$, then $\chi(G) = \sum_{i=1}^{k} \chi(G_i)$.

One of the most useful lower bounds for the chromatic number of a graph is stated next.

Proposition 1.2.3. *If H is a subgraph of a graph G, then $\chi(H) \leq \chi(G)$.*

The *clique number* $\omega(G)$ of a graph G is the maximum order of a complete subgraph of G. The following result is therefore a consequence of Proposition 1.2.3.

Corollary 1.2.4. *For every graph G, $\omega(G) \leq \chi(G)$.*

By Corollary 1.2.4 (or, in fact, by Observation 1.2.1(c)), if a graph G contains a triangle, then $\chi(G) \geq 3$. There are graphs G for which $\chi(G)$ and $\omega(G)$ may differ significantly.

As far as upper bounds for the chromatic number of a graph are concerned, the following result gives such a bound in terms of the maximum degree of the graph.

Theorem 1.2.5. *For every graph G, $\chi(G) \leq \Delta(G) + 1$.*

For each positive integer n, $\chi(K_n) = n = \Delta(K_n) + 1$ and for each odd integer $n \geq 3$, $\chi(C_n) = 3 = \Delta(C_n) + 1$. The British mathematician Rowland Leonard Brooks showed that these two classes of graphs are the only connected graphs with this property.

Theorem 1.2.6 ([11]). *If G is a connected graph that is neither an odd cycle nor a complete graph, then $\chi(G) \leq \Delta(G)$.*

The *distance* $d(u, v)$ between two vertices u and v in a connected graph G is the length of a shortest $u-v$ path and the *diameter* diam(G) of G is the greatest distance between two vertices of G. For a vertex v in a connected graph G, the *eccentricity* $e(v)$ of v is the greatest distance between v and a vertex in G. Thus, the diameter of G is also the largest eccentricity among all vertices of G. There is an upper bound for the chromatic number of a connected graph in terms of the order and diameter of the graph, which is due to Vašek Chvátal.

Theorem 1.2.7 ([29]). *If G is a connected graph of order n and diameter d, then*

$$\chi(G) \leq n - d + 1.$$

1.3   Proper Edge Colorings

A *proper edge coloring* c of a nonempty graph G is a function $c : E(G) \rightarrow S$, where S is a set of colors (and typically $S = [k]$ or $S = \mathbb{Z}_k$ for some integer $k \geq 2$), with the property that $c(e) \neq c(f)$ for every two adjacent edges e and f of G. If the colors are chosen from a set of k colors, then c is called a *k-edge coloring* of G. The minimum positive integer k for which G has a k-edge coloring is called the *chromatic index* of G and is denoted by $\chi'(G)$.

It is immediate for every nonempty graph G that $\chi'(G) \geq \Delta(G)$. The most important theorem dealing with chromatic index is one obtained by the Russian mathematician Vadim Vizing.

Theorem 1.3.1 ([74]). *For every nonempty graph G,*

$$\chi'(G) \leq \Delta(G) + 1.$$

As a result of Vizing's theorem, the chromatic index of every nonempty graph G is one of two numbers, namely $\Delta(G)$ or $\Delta(G) + 1$. A graph G with $\chi'(G) = \Delta(G)$ is called a *class one graph* while a graph G with $\chi'(G) = \Delta(G) + 1$ is called a *class two graph* . The chromatic index of complete graphs is given in the following result.

Theorem 1.3.2. *For each integer $n \geq 2$,*

$$\chi'(K_n) = \begin{cases} n - 1 & \textit{if n is even} \\ n & \textit{if n is odd.} \end{cases}$$

Therefore, K_n is a class one graph if n is even and is a class two graph if n is odd. The fact that K_n is a class one graph if and only if n is even is also a consequence of the following.

Theorem 1.3.3. *A regular graph G is a class one graph if and only if G is* 1-*factorable.*

An immediate consequence of this result is stated next.

Corollary 1.3.4. *Every regular graph of odd order is a class two graph.*

The next two results describe classes of graphs that are class one graphs. The first theorem is due to Denés König.

Theorem 1.3.5 ([48]). *Every bipartite graph is a class one graph.*

If a graph G of odd order has sufficiently many edges, then G must be a class two graph. A graph G of order n and size m is called *overfull* if $m > \Delta(G)\lfloor n/2 \rfloor$. If G has even order n, then $m \leq \Delta(G)\lfloor n/2 \rfloor$ and so G is not overfull. On the other hand, a graph of odd order may be overfull.

Theorem 1.3.6. *Every overfull graph is a class two graph.*

1.4   Eulerian Graphs and Digraphs

A circuit in a nontrivial connected graph G that contains every edge of G (necessarily exactly once) is an *Eulerian circuit in a graph*. A connected graph G is *Eulerian* if G contains an Eulerian circuit. The following characterization of Eulerian graphs is attributed to Leonhard Euler.

Theorem 1.4.1 ([31]). *A nontrivial connected graph G is Eulerian if and only if every vertex of G has even degree.*

These concepts dealing with graphs have analogues for digraphs as well. An *Eulerian circuit* in a connected digraph D is a directed circuit that contains every arc of D. A connected digraph that contains an Eulerian circuit is an *Eulerian digraph*. As with the characterization of Eulerian graphs in Theorem 1.4.1, the characterization of Eulerian digraphs stated next is given in terms of the digraph analogues of 'degrees'. The *outdegree* od v of a vertex v in a digraph D is the number of arcs incident with and directed away from v, while the *indegree* id v of v is the number of arcs incident with and directed towards v.

Theorem 1.4.2. *Let D be a nontrivial connected digraph. Then D is Eulerian if and only if* od $v =$ id v *for every vertex v of D.*

For example, since od $v =$ id v for every vertex v of the digraph D of Fig. 1.1, it is Eulerian and $(v_1, v_2, v_3, v_7, v_6, v_2, v_4, v_6, v_5, v_4, v_1)$ is an Eulerian circuit of D.

1.5   A Theorem from Discrete Mathematics

While we will use some basic facts and results from discrete mathematics throughout our discussion, there is one result that we will encounter so often that it is valuable to state here at the beginning. This result deals with combinations with repetition.

Theorem 1.5.1. *Let A be a multiset containing ℓ different kinds of elements, where there are at least r elements of each kind. The number of different selections of r elements from A is $\binom{r+\ell-1}{r}$.*

Fig. 1.1 An Eulerian digraph D

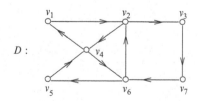

To illustrate Theorem 1.5.1, suppose that A is a multiset containing four different kinds of elements, say $1, 2, 3, 4$, and there are at least three elements of each kind. For example, perhaps,

$$A = \{1, 1, 1, 2, 2, 2, 3, 3, 3, 4, 4, 4\}.$$

Hence, $\ell = 4$ and $r = 3$. By Theorem 1.5.1, the number of selections of $r = 3$ elements from A is $\binom{r+\ell-1}{r} = \binom{3+4-1}{3} = \binom{6}{3} = 20$. The 20 selections here are the following multisets:

$$
\begin{aligned}
&A_1 = \{1, 1, 1\}, \ A_2 = \{2, 2, 2\}, \ A_3 = \{3, 3, 3\}, \ A_4 = \{4, 4, 4\} \\
&A_5 = \{1, 1, 2\}, \ A_6 = \{1, 1, 3\}, \ A_7 = \{1, 1, 4\}, \ A_8 = \{2, 2, 1\} \\
&A_9 = \{2, 2, 3\}, \ A_{10} = \{2, 2, 4\}, \ A_{11} = \{3, 3, 1\}, \ A_{12} = \{3, 3, 2\} \\
&A_{13} = \{3, 3, 4\}, \ A_{14} = \{4, 4, 1\}, \ A_{15} = \{4, 4, 2\}, \ A_{16} = \{4, 4, 3\} \\
&A_{17} = \{1, 2, 3\}, \ A_{18} = \{1, 2, 4\}, \ A_{19} = \{1, 3, 4\}, \ A_{20} = \{2, 3, 4\}.
\end{aligned}
$$

Chapter 2
Binomial Edge Colorings

In [76], a number of edge colorings were described that gave rise to various vertex colorings of interest. In one instance, the color of a vertex was defined as the set of colors of the edges incident with the vertex, with the goal to minimize the number of colors so that the resulting coloring is vertex-distinguishing. In this chapter, we consider edge colorings that result in the same vertex coloring, but with a different goal in mind. Here, for a fixed number k of edge colors $1, 2, \ldots, k$, we wish to determine graphs of minimum order having the property that for every subset S of $[k] = \{1, 2, \ldots, k\}$, there is a vertex whose color is S.

2.1 Strong Edge Colorings

A vertex coloring of a graph G is *vertex-distinguishing* if no two vertices of G are assigned the same color. An edge coloring c of a graph G has been referred to as a *strong edge coloring* of G if c is a proper edge coloring that induces a vertex-distinguishing coloring which assigns to each vertex v of G the set of colors of the edges incident with v. Such an edge coloring is also called vertex-distinguishing. Since no two vertices of G are colored the same, no two vertices are assigned the same set. Consequently, for every two vertices of G, there is an edge incident with one of these two vertices whose color is not assigned to any edge incident with the other vertex. The minimum positive integer k for which G has a strong k-edge coloring has been called the *strong chromatic index* of G, denoted by $\chi'_s(G)$. Since every strong edge coloring of a nonempty graph G is a proper edge coloring of G, it follows that $\Delta(G) \leq \chi'(G) \leq \chi'_s(G)$. The concept of strong edge colorings of graphs was introduced independently by Aigner et al. [1], Černý et al. [13], Horňá and Soták [46, 47] and Burris and Schelp [12]. The terms *strong edge coloring* and *strong chromatic index* were introduced in [12, 32].

© The Author 2016
P. Zhang, *A Kaleidoscopic View of Graph Colorings*, SpringerBriefs in Mathematics,
DOI 10.1007/978-3-319-30518-9_2

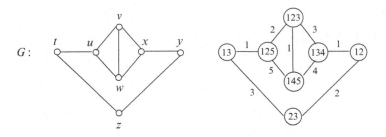

Fig. 2.1 A strong 5-edge coloring of a graph G

In order to illustrate this type of edge coloring, we determine the strong chromatic index of the graph G of Fig. 2.1. Since $\chi'(G) = 3$, it follows that $\chi'_s(G) \geq 3$. However, $\chi'_s(G) \neq 3$ since any proper 3-edge coloring of G must assign the color $\{1, 2, 3\}$ to every vertex of degree 3. Moreover, if $\chi'_s(G) = 3$, then since the order of G is 7 the seven vertices of G would have to be colored with the seven nonempty sets of $\{1, 2, 3\}$. Since the minimum degree of G is $\delta(G) = 2$, no vertex of G can be colored $\{1\}$, $\{2\}$ or $\{3\}$. Furthermore, $\chi'_s(G) \neq 4$, for suppose that there is a strong 4-edge coloring c of G. We may assume that $c(vw) = 1$. Since c is a proper edge coloring, none of the edges uv, uw, vx and wx can be colored 1. Hence, two of these four edges must be assigned the same color and the remaining two edges must be assigned different colors, say uv and wx are colored 2. Thus, all of the vertices u, v, w and x are assigned a color that is a 3-element set containing 2. This, however, implies that two of these vertices are colored the same, which is impossible. Hence, $\chi'_s(G) \geq 5$. The strong 5-edge coloring of G in Fig. 2.1 shows that $\chi'_s(G) = 5$. where the sets $\{a\}$, $\{a, b\}$ and $\{a, b, c\}$ are denoted by a, ab, abc, respectively, with $a < b < c$.

The argument used to verify that the strong chromatic index of the graph G of Fig. 2.1 is 5 suggests a more general observation. If a graph G has strong chromatic index k, say, then the induced color assigned to a vertex of degree r is one of the r-element subsets of $\{1, 2, \ldots, k\}$.

Observation 2.1.1. *If a graph G of order at least 3 contains more than $\binom{k}{r}$ vertices of degree r ($1 \leq r \leq \Delta(G)$) for some positive integer k, then $\chi'_s(G) \geq k + 1$.*

Although $\Delta(G) + 1$ is an upper bound for $\chi'(G)$ by Vizing's Theorem, $\Delta(G) + 1$ is not an upper bound for $\chi'_s(G)$, as the graph of Fig. 2.1 shows. In fact, there is no constant c such that $\chi'_s(G) \leq \Delta(G) + c$ for every graph G since, for example, if $n = \binom{\ell+1}{2} + 1$, then

$$\chi'_s(C_n) \geq \ell + 2 = \Delta(C_n) + \ell$$

by Observation 2.1.1. The following sharp upper bound for the strong chromatic index was obtained by Bazgan, Harkat-Benhamdine, Li and Woźniak, verifying a conjecture by Burris and Schelp (see [5]).

Theorem 2.1.2. *If G is a connected graph of order $n \geq 3$, then $\chi'_s(G) \leq n + 1$.*

This topic has also been discussed in many research papers (see [1, 6, 12, 13, 15, 32, 46, 47, 76], for example).

2.2 Proper k-Binomial-Colorable Graphs

If G is a graph of order n with strong chromatic index k, then $n \leq 2^k$ since there are 2^k subsets of $[k]$. Furthermore, if $\chi'_s(G) = k$ and G has order 2^k, then G must contain exactly $\binom{k}{r}$ vertices of degree r for every integer r with $0 \leq r \leq k$. For example, the graph G of order $16 = 2^4$ in Fig. 2.2 has $\binom{4}{r}$ vertices of degree r for every integer r with $0 \leq r \leq 4$. The proper 4-edge coloring of G in Fig. 2.2 has the property that $\{c'(v) : v \in V(G)\} = \mathcal{P}([4])$ and so $\chi'_s(G) = 4$.

For an integer $k \geq 2$, a *k-binomial graph* is a graph containing $\binom{k}{r}$ vertices of degree r for each integer r with $0 \leq r \leq k$. (There is no 1-binomial graph.) Thus, such a graph G has order

$$n = \sum_{r=0}^{k} \binom{k}{r} = 2^k$$

and size

$$m = \frac{1}{2} \sum_{r=0}^{k} r \binom{k}{r} = k2^{k-2}.$$

A graph G is a *binomial graph* if G is a k-binomial graph for some integer $k \geq 2$. These concepts were introduced and studied in [27].

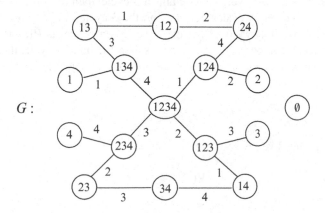

Fig. 2.2 A graph of order 16 with strong chromatic index 4

Fig. 2.3 Eight k-binomial
graphs for $k = 2, 3$

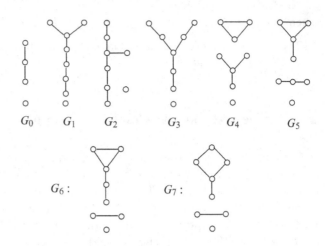

$$G_0 \qquad G_1 \qquad G_2 \qquad G_3 \qquad G_4 \qquad G_5$$

$$G_6 : \qquad\qquad G_7 :$$

The unique 2-binomial graph G_0 and the seven 3-binomial graphs $G_1 - G_7$ are shown in Fig. 2.3. Let's see why $G_1 - G_7$ are the only 3-binomial graphs. Since the degree sequence of a 3-binomial graph G is 3, 2, 2, 2, 1, 1, 1, 0, its size is 6. Let H be the subgraph of G not containing the isolated vertex. If H is a tree, then H must be obtained by three subdivisions of $K_{1,3}$. This can be done in three ways: (1) subdividing each edge of $K_{1,3}$ once, (2) subdividing one edge of $K_{1,3}$ twice and one edge once and (3) subdividing one edge of $K_{1,3}$ three times. Thus, there are three such 3-binomial graphs. If H is not a tree, then H must be disconnected where each component contains at least two vertices. Since the maximum size of such a graph having three components is 5, it follows that H has exactly two components, one of order ℓ, say, and the other of order $7 - \ell$, where $2 \le \ell \le 5$. Since the size of H is 6 and the minimum size of the two components is $(\ell - 1) + (6 - \ell) = 5$, one component of H must be a tree and the other a unicyclic graph (and so contains exactly one cycle). Because H has three end-vertices, H does not contain a 5-cycle. If the unicyclic component of H is a 3-cycle, then G is G_4. If the unicyclic component of H is not a 3-cycle but contains a 3-cycle, then (a) this component has order 4 or 5, (b) the vertex of degree 3 lies on the 3-cycle in H and (c) the acyclic component of H is either P_3 or P_2. If the acyclic component is P_3, then G is G_5; while if the acyclic component is P_2, then G is G_6. If H has a 4-cycle, then G is G_7.

For an integer $k \ge 2$, a *proper k-binomial-coloring* of a graph G is a proper edge coloring

$$c : E(G) \to [k] = \{1, 2, \ldots, k\}$$

such that the induced vertex coloring

$$c' : V(G) \to \mathscr{P}([k]),$$

where $c'(v)$ is the set of colors of the edges incident with v, is both vertex-distinguishing and satisfies the condition that

$$\{c'(v) : \; v \in V(G)\} = \mathscr{P}([k]).$$

A graph G admitting a proper k-binomial-coloring is a *proper k-binomial-colorable graph*. Necessarily, a proper k-binomial-colorable graph is a k-binomial graph. A graph G is a *proper binomial-colorable graph* if G is a proper k-binomial-colorable for some integer k. Each of the graphs $G_0, G_1, \ldots, G_5$ in Fig. 2.3 is a proper binomial-colorable graph. A proper 3-binomial-coloring of each of these graphs is shown in Fig. 2.4. Since no graph containing K_2 as a component can be a proper binomial-colorable graph, the graphs G_6 and G_7 (in Figs. 2.3 and 2.4) are not proper binomial-colorable graphs. Furthermore, the graph of Fig. 2.2 is a proper 4-binomial-colorable graph. Note that for each integer $k \geq 2$, in a proper k-binomial-coloring of a k-binomial graph, each color in $[k]$ is assigned to exactly 2^{k-2} edges.

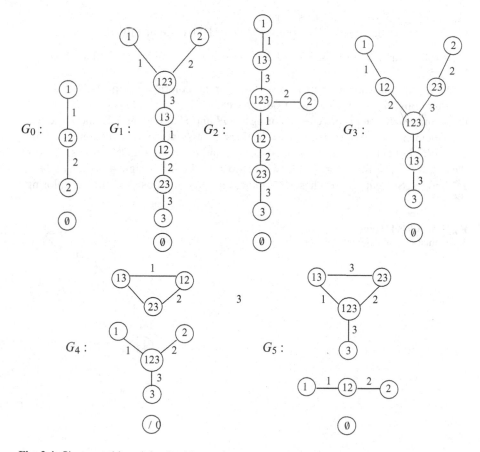

Fig. 2.4 Six proper binomial-colorable graphs

This brings up the following the following conjecture.

Conjecture 2.2.1. *If G is a k-binomial graph for some integer $k \geq 2$ that does not contain K_2 as a component, then G has a proper binomial coloring using colors from the set $[k]$.*

We saw in Figs. 2.3 and 2.4 that there is a k-binomial-colorable graph for $k \in \{2, 3, 4\}$. Much more can be said.

Theorem 2.2.2 ([27]). *For every integer $k \geq 2$, there exists a proper k-binomial-colorable graph.*

Proof. We proceed by induction on k. We have already seen that there exists a proper 2-binomial-colorable graph and a proper 3-binomial-colorable graph. Assume that there exists a proper k-binomial-colorable graph for some integer $k \geq 3$. We show that there exists a proper $(k + 1)$-binomial-colorable graph.

By the induction hypothesis, there exists a proper k-binomial-colorable graph H for some integer $k \geq 3$. Let c be a proper k-binomial coloring of H. Express $V(H)$ as $\{v_1, v_2, \ldots, v_{2^k}\}$ such that the following hold:

(1) For $1 \leq i \leq j \leq 2^k$, $\deg v_i \leq \deg v_j$.
(2) If $\deg v_i = \deg v_j$ and $c'(v_i)$ precedes $c'(v_j)$ lexicographically, then $i < j$.

So, for the graph G_3 in Figs. 2.3 and 2.4, the vertices of G_3 are labeled as shown in Fig. 2.5.

Next, we construct a proper $(k + 1)$-binomial-colorable graph G. Let H' be another copy of the graph H where the vertex v_i ($1 \leq i \leq 2^k$) in H is labeled v'_i in H' and where the proper edge colorings of H and H' are identical. Therefore, for each element S of $\mathscr{P}([k])$, exactly two vertices are assigned the color S, one in H and one in H'. Let G be the graph obtained from H and H' by adding 2^{k-1} edges, namely the edges $v_{2i-1}v'_{2i}$ for $1 \leq i \leq 2^{k-1}$, where the color $k+1$ is assigned to each of these 2^{k-1} edges. Since no two of these 2^{k-1} edges are adjacent, this $(k + 1)$-edge coloring

Fig. 2.5 A labeled proper 3-binomial-colorable graph

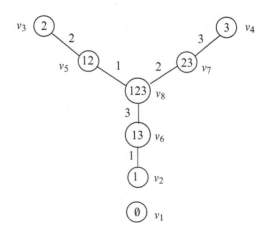

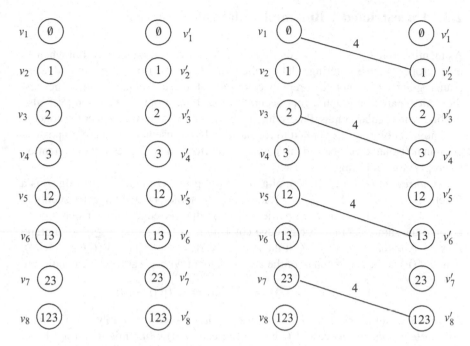

Fig. 2.6 Illustrating a step of the proof of Theorem 2.2.2

of G is proper. Furthermore, for each element S of $\mathscr{P}([k])$, exactly one vertex of G is colored S and exactly one vertex is colored $S \cup \{k + 1\}$, that is, for each element S' of $\mathscr{P}([k + 1])$, exactly one vertex of G is colored S'. This is illustrated in Fig. 2.6 for $k = 3$.

The number of vertices of degree r in G for $0 \le r \le k$ is therefore the sum of the number $\binom{k}{r-1}$ of vertices of degree $r - 1$ in H and the number $\binom{k}{r}$ of vertices of degree r in H'. Since

$$\binom{k}{r-1} + \binom{k}{r} = \binom{k+1}{r},$$

it follows that G has $\binom{k+1}{r}$ vertices of degree r. Therefore, G is a proper $(k + 1)$-binomial-colorable graph. By the Principle of Mathematical Induction, there exists a proper k-binomial-colorable graph for every integer $k \ge 2$. $\square$

2.3   Unrestricted *k*-Binomial-Colorable Graphs

As mentioned in Chap. 1, edge colorings of graphs were introduced by Tait when he
used proper 3-edge colorings (later called *Tait colorings*) of 3-regular bridgeless
planar graphs to generate 4-region colorings of these graphs. In fact, the first
theoretical paper on graph theory occurred in an 1891 article of Petersen [59], also
dealing with regular graphs. Proceeding in this manner, we now consider the concept
of binomial colorings as applied to regular graphs. Consequently, these graphs can
contain no isolated vertices. In this case, we therefore no longer restrict our attention
to proper edge colorings.

   An *unrestricted edge coloring* of a nonempty graph G (a graph with edges) is a
function $c : E(G) \rightarrow S = [k] = \{1, 2, \ldots, k\}$ for some positive integer k such that
no condition is placed on c. In particular, two adjacent edges may or may not be
colored the same. An unrestricted edge coloring $c : E(G) \rightarrow [k]$, $k \geq 2$, of a graph
G is a *k-binomial coloring* of G if the induced vertex coloring $c' : V(G) \rightarrow \mathscr{P}([k])$,
where $c'(v)$ is the set of colors of the edges incident with v, satisfies the condition

$$\{c'(v) : \ v \in V(G)\} = \mathscr{P}^*([k]) = \mathscr{P}([k]) - \{\emptyset\}.$$

A graph G is an *unrestricted k-binomial-colorable graph* (or simply a *k-binomial-
colorable graph*) in this case if G has an (unrestricted) k-binomial-coloring. Here,
we are interested in, for a fixed integer $k \geq 2$, the existence of an r-regular
k-binomial-colorable graph of order n, where r and/or n is as small as possible.
Necessarily, $r \geq k$ and $n \geq 2^k - 1$. These concepts were introduced and studied
in [27]. We begin with the following result.

Theorem 2.3.1 ([27]). *For each integer $k \geq 2$, there exists a k-regular k-binomial-
colorable graph of order 2^k.*

Proof. We show, in fact, for each integer $k \geq 2$, that the k-regular k-cube $G = Q_k$
of order 2^k is a k-binomial-colorable graph. The vertices of G can be labeled by the
set of k-bit sequences. For each integer ℓ with $0 \leq \ell \leq k$, let V_ℓ be the set consisting
of the $\binom{k}{\ell}$ vertices of G whose k-bit labels have exactly ℓ terms equal to 1. Thus,
$V(G) = \bigcup_{\ell=0}^{k} V_\ell$. For $0 \leq \ell \leq k$, let

$$V_\ell = \left\{ v_{\ell,1}, v_{\ell,2}, \ldots, v_{\ell,\binom{k}{\ell}} \right\},$$

where the $\binom{k}{\ell}$ vertices in each set V_ℓ are listed so that their labels are in reverse
lexicographical order. For example, for $k = 3$, it follows that

$$V_0 = \{v_{0,1}\}, V_1 = \{v_{1,1}, v_{1,2}, v_{1,3}\}, V_2 = \{v_{2,1}, v_{2,2}, v_{2,3}\}, V_3 = \{v_{3,1}\},$$

where

$$v_{0,1} = (0,0,0), v_{1,1} = (1,0,0), v_{1,2} = (0,1,0), v_{1,3} = (0,0,1),$$
$$v_{2,1} = (1,1,0), v_{2,2} = (1,0,1), v_{2,3} = (0,1,1), v_{3,1} = (1,1,1).$$

Since two vertices u and v are adjacent in G if and only if the labels of u and v differ in exactly one position, it follows that one of u and v belongs to V_i and the other belongs to V_{i+1} for some i ($0 \leq i \leq k-1$), say $u \in V_i$ and $v \in V_{i+1}$. So, every term having the value 1 for u also has the value 1 for v.

It remains to show that G has an unrestricted edge coloring $c : E(G) \to [k]$ such that $\{c'(v) : v \in V(G)\} = \mathscr{P}^*([k])$. First, define

$$c(v_{0,1}v_{1,i}) = i \text{ for } i = 1, 2, \ldots, \binom{k}{1} = k.$$

Then $c'(v_{0,1}) = [k]$. Next, define $c(e_i) = i$ for each edge e_i incident with $v_{1,i}$ and so $c'(v_{1,i}) = \{i\}$ for $1 \leq i \leq k$.

Assume, for a fixed integer j with $2 \leq j \leq k-1$ and for all integers i with $2 \leq i \leq j$, that all edges joining a vertex in V_{i-1} and a vertex in V_i have been assigned colors by the coloring c so that $c'(v_{i-1,t})$, $1 \leq t \leq \binom{k}{i-1}$, is the subset of $[k]$ in which $s \in c'(v_{i-1,t})$ if and only if the sth coordinate of $v_{i-1,t}$ is 1. Furthermore, assume that this is the case for $c'(v_{j,t})$ as well, where $1 \leq t \leq \binom{k}{j}$, taking into consideration only those edges joining $v_{j,t}$ to the vertices in V_{j-1}.

Next, let $x \in V_j$ and $y \in V_{j+1}$ such that $xy \in E(G)$. The labels of x and y therefore differ in exactly one coordinate. Let $x = (x_1, x_2, \ldots, x_k)$ and $y = (y_1, y_2, \ldots, y_k)$, where then exactly j of the coordinates of x have the value 1, exactly $j+1$ of the coordinates of y have the value 1 and there is a unique integer r with $1 \leq r \leq k$ such that $x_r = 0$ and $y_r = 1$. If p is the largest integer, $1 \leq p \leq k$, such that $p < r$ and $x_p = y_p = 1$, then define $c(xy) = p$. If $r = 1$ or $r \geq 2$ and $x_i - y_i = 0$ for $1 \leq i \leq r-1$, then p is the largest integer for which $x_p = y_p = 1$. The coloring c is illustrated in Fig. 2.7 for $k = 4$ where each k-bit $(x_1, x_2, \ldots, x_k)$ is denoted by $x_1x_2 \ldots x_k$.

It remains to show that for each vertex in $V_j \cup V_{j+1}$ where $j \leq k-1$, the induced color of the vertex consists of the subscripts of those terms having value 1. First, let $x \in V_j$. Then $x = (x_1, x_2, \ldots, x_k)$ and exactly j of the terms $x_1, x_2, \ldots, x_k$ are 1. Suppose that these j terms are $x_{n_1}, x_{n_2}, \ldots, x_{n_j}$ where $1 \leq n_1 < n_2 < \cdots < n_j$. Then the set of colors of the edges joining x with the vertices in V_{j-1} is $\{n_1, n_2, \ldots, n_j\}$. Since the color of any edge joining x and a vertex $y = (y_1, y_2, \ldots, y_k) \in V_{j+1}$ is some integer n_t for which $x_{n_t} = y_{n_t} = 1$, it follows that $c'(x) = \{n_1, n_2, \ldots, n_j\}$. Next, let

$$y = (y_1, y_2, \ldots, y_k) \in V_{j+1}$$

and exactly $j+1$ of the terms $y_1, y_2, \ldots, y_k$ are 1. Suppose that these $j+1$ terms are $y_{m_1}, y_{m_2}, \ldots, y_{m_{j+1}}$, where $1 \leq m_1 < m_2 < \cdots < m_{j+1} \leq k$. By the defining property of c, each edge joining y and a vertex in V_j is colored with some integer in $\{m_1, m_2, \ldots, m_{j+1}\}$. We now show that for each m_i with $1 \leq i \leq j+1$, there is an edge joining y and a vertex $x \in V_j$ that is colored m_i by c. Let

$$x = (y_1, y_2, \ldots, y_{m_i}, \ldots, y_{m_{i+1}} - 1, \ldots, y_k).$$

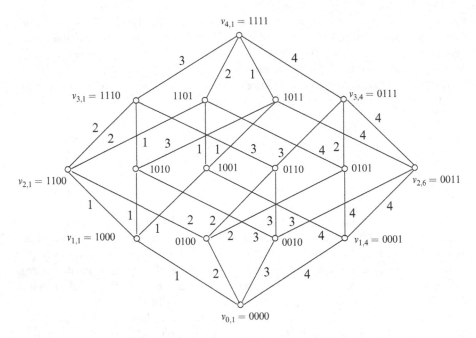

Fig. 2.7 Illustrating the coloring c in the proof of Theorem 2.3.1 for $k = 4$

Then x has exactly j terms having value 1 and so $x \in V_j$. The labels of x and y differ in exactly one position, namely the m_{i+1}th position, where $x_{m_{i+1}} = y_{m_{i+1}} - 1$. Since m_i is the largest integer such that $x_{m_i} = y_{m_i} = 1$, it follows that $c(xy) = m_i$. Hence, $c'(y) = \{m_1, m_2, \ldots, m_{j+1}\}$, taking into consideration only those edges joining y to the vertices in V_j. Therefore, c is a k-binomial coloring of G and so G is a k-binomial-colorable graph. $\square$

The following is a consequence of the proof of Theorem 2.3.1.

Corollary 2.3.2 ([27]). *For each even integer $k \geq 4$, there exists a k-regular k-binomial-colorable graph of order $2^k - 1$.*

Proof. Let G be the graph of order $2^k - 1$ obtained from the graph $Q_k - v_{k,1}$ described in the proof of Theorem 2.3.1 by adding the $k/2$ edges $v_{k-1,i}v_{k-1,i+1}$ for each odd integer i with $1 \leq i \leq k - 1$. Let c be the coloring of Q_k defined in the proof of Theorem 2.3.1. Note that for each odd integer i with $1 \leq i \leq k - 1$, the sets $c'(v_{k-1,i})$ and $c'(v_{k-1,i+1})$ are both $(k - 1)$-element subsets of the k-element set $[k]$ and so $|c'(v_{k-1,i}) \cap c'(v_{k-1,i+1})| \geq k - 2 \geq 2$. Now define the edge coloring c^* of G by $c^*(e) = c(e)$ if $e \in E(Q_k - v_{k,1})$ and $c^*(v_{k-1,i}v_{k-1,i+1}) = s_i$ where $s_i \in c'(v_{k-1,i}) \cap c'(v_{k-1,i+1})$. For each $v \in V(G)$, the vertex color of v induced by c^* is in fact $c'(v)$. Hence, c^* is a k-binomial coloring of G and so G is a k-binomial-colorable graph. $\square$

While there cannot exist a k-regular k-binomial-colorable graph of order $2^k - 1$ for any odd integer k, there is, however, a $(k+1)$-regular k-binomial-colorable graph of order $2^k - 1$ for every odd integer $k \geq 5$. For a graph G and two disjoint subsets X and Y of $V(G)$, let $[X, Y]$ denote the set of edges joining a vertex of X and a vertex of Y.

Theorem 2.3.3 ([27]). *For each odd integer $k \geq 5$, there exists a $(k + 1)$-regular k-binomial-colorable graph of order $2^k - 1$.*

Proof. We begin with the k-cube Q_k having the edge coloring c described in the proof of Theorem 2.3.1. As in the proof of Theorem 2.3.1, the vertices of Q_k are labeled by the set of k-bit sequences. For $0 \leq \ell \leq k$, let V_ℓ be the set of the $\binom{k}{\ell}$ vertices of Q_k whose k-bit labels have exactly ℓ terms equal to 1. Thus, each set V_ℓ is an independent set of vertices in Q_k. Hence, $V(Q_k) = \bigcup_{\ell=0}^{k} V_\ell$. For $0 \leq \ell \leq k$, let

$$V_\ell = \left\{ v_{\ell,1}, v_{\ell,2}, \ldots, v_{\ell,\binom{k}{\ell}} \right\},$$

where the vertices of each set V_ℓ are listed so that their labels are in reverse lexicographical order. For a vertex $v = (a_1, a_2, \ldots, a_k)$ in Q_k, the vertex

$$\overline{v} = (1 - a_1, 1 - a_2, \ldots, 1 - a_k)$$

is the complementary vertex of v. If $v \in V_\ell$, then $\overline{v} \in V_{k-\ell}$.

Let $F = Q_k - v_{k,1}$ be the graph of order $2^k - 1$ having $2^k - k - 1$ vertices of degree k and k vertices of degree $k - 1$. From F, our goal is to construct a $(k + 1)$-regular k-binomial-colorable graph G of order $2^k - 1$ where $V(G) = V(F)$ and $E(F) \subseteq E(G)$. We will then define an edge coloring c_0 on G such that $c_0(e) = c(e)$ if $e \in E(F)$ and $c_0'(v) = c'(v)$ for each $v \in V(G)$. For this purpose, it is useful to consider the diagram in Fig. 2.8 showing the general structure of the graph G to be constructed. We now describe the construction of this graph (as well as the notation in Fig. 2.8) as follows.

Let $k = 2p + 1$, where then $p \geq 2$. For $1 \leq \ell \leq p - 1$, let H_ℓ be the $\left(\binom{k}{\ell} - 1 \right)$-regular bipartite subgraph of the complement $\overline{F}$ of F with partite sets V_ℓ and $V_{k-\ell}$ containing none of the edges $v\overline{v}$, where $v \in V_\ell$. Since H_ℓ is a regular bipartite graph, it contains a perfect matching M_ℓ. All the matchings M_ℓ ($1 \leq \ell \leq p - 1$) are then added to F. For each edge xy in M_ℓ, there is at least one term i in the k-bit sequences of x and y that both equal 1 (since x and y are not complementary vertices). In this case, the edge xy is colored i by c_0, resulting in $c_0'(x) = c'(x)$ and $c_0'(y) = c'(y)$.

The subgraph $F_p = F[V_p \cup V_{p+1}]$ is a $(p + 1)$-regular bipartite subgraph of F with partite sets V_p and V_{p+1}, where $|V_p| = |V_{p+1}| = \binom{2p+1}{p}$. Let H_p be the bipartite subgraph of $\overline{F}$ with partite sets V_p and V_{p+1} having edge set

$$E(H_p) = [V_p, V_{p+1}] - E(F_p) - \{v\overline{v} : v \in V_p\}.$$

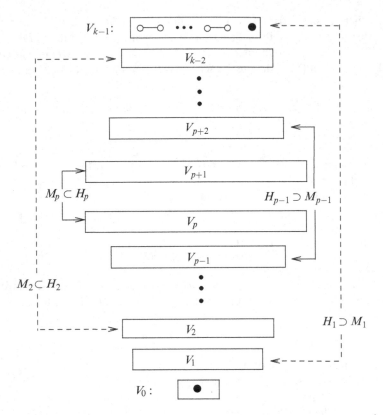

Fig. 2.8 The structure of a $(k + 1)$-regular k-binomial-colorable graph G of order $2^k - 1$ in Theorem 2.3.3

Since H_p is a $\left(\binom{2p+1}{p} - p - 2\right)$-regular bipartite graph, H_p has a perfect matching M_p. This matching is also added to F. Each edge $xy \in M_p$ has at least one term i in the k-bit sequences of x and y that both equal 1. The edge xy is colored i by c_0.

At this stage, each vertex of the graph currently constructed has degree $k + 1$ except for each of the $k = 2p + 1$ vertices in V_{k-1} and the single vertex in V_0 has degree k.

To complete the construction of G, we add $p + 1$ edges to F, namely (1) the p edges $v_{k-1,2i-1}v_{k-1,2i}$ for $i = 1, 2, \ldots, p$ and (2) the edge $v_{0,1}v_{k-1,k}$. For each of the p added edges xy described in (1), there is at least one term i in the k-bit sequences of x and y that both equal 1. The edge xy is colored i by c_0 and the edge $v_{0,1}v_{k-1,k}$ is colored k by c_0. Thus, $c_0'(x) = c'(x)$ and $c_0'(y) = c'(y)$ for all these added edges xy in G. The vertices $v_{0,1}$ and $v_{k-1,k}$ are drawn as solid vertices. Consequently, c_0 is a k-binomial coloring of G and so G is a $(k + 1)$-regular k-binomial-colorable graph of order $2^k - 1$.   □

Chapter 3
Kaleidoscopic Edge Colorings

In this chapter, we consider an edge coloring problem in graphs that can be used to model certain situations, one of which we now describe. Suppose that a hard-line network of n computers is to be constructed. Each of these computers requires k different types of connections. There are r locations on the back of each computer at which ports can be placed. Each computer needs to have at least one connection of each type and, for security reasons, no two computers can have more than one connection between them. In order to maximize the number of fail-safe connections, every port is to be used. Furthermore, it is advantageous for a computer technician to be able to distinguish the computers based only on the number of types of connections they have. For which values of n, k and r is such a situation possible?

3.1 Introduction

A well-known observation in graph theory concerning the degrees of the vertices of a graph is that every nontrivial graph contains at least two vertices having the same degree. Indeed, it is known that for every integer $n \geq 2$, there are exactly two graphs of order n having exactly two vertices of the same degree and these two graphs are complements of each other. Consequently, in any decomposition of the complete graph K_n of order n into two graphs, necessarily into a graph G and its complement $\overline{G}$, there are at least two vertices u and v such that $\deg_G u = \deg_G v$ (and so $\deg_{\overline{G}} u = \deg_{\overline{G}} v$ as well). In particular, for every decomposition of a complete graph K_n into two graphs G_1 and G_2 (where then $G_2 = \overline{G}_1$) such that each vertex of K_n is incident with at least one edge in each of G_1 and G_2, there is associated with each vertex v of K_n an ordered pair (a, b) of positive integers with $a = \deg_{G_1} v$ and $b = \deg_{G_2} v$. Consequently, for each such decomposition of K_n, there are at least two vertices with the same ordered pair. In fact, this is not only true of decompositions of the complete graph into two graphs but decompositions of every regular graph

© The Author 2016
P. Zhang, *A Kaleidoscopic View of Graph Colorings*, SpringerBriefs in Mathematics,
DOI 10.1007/978-3-319-30518-9_3

into two graphs. Indeed, for a given regular graph G, there is a question of whether there exists a decomposition of G into $k \geq 3$ graphs $G_1, G_2, \ldots, G_k$ such that (1) each vertex of G is incident with at least one edge of every graph G_i and (2) for every two vertices u and v of G, $\deg_{G_i} u \neq \deg_{G_i} v$ for some i. By assigning the color i ($1 \leq i \leq k$) to each edge of G_i, we are led to the following graph coloring concept, first introduced in [28].

For an r-regular graph G, let $c : E(G) \rightarrow [k] = \{1, 2, \ldots, k\}$, $k \geq 3$, be an edge coloring of G, where every vertex of G is incident with at least one edge of each color. Thus, $r \geq k$. For a vertex v of G, the *set-color* $c_s(v)$ of v is defined as the set of colors of the edges incident with v. Thus, $c_s(v) = [k]$ for every vertex v of G. That is, each such edge coloring of G induces a *set-regular* vertex coloring of G. The *multiset-color* $c_m(v)$ of v is defined as the ordered k-tuple $(a_1, a_2, \ldots, a_k)$ or $a_1 a_2 \ldots a_k$, where a_i ($1 \leq i \leq k$) is the number of edges in G colored i that are incident with v. Hence, each a_i is a positive integer and $\sum_{i=1}^{k} a_i = r$. Such an edge coloring c is called a k-*kaleidoscopic coloring* of G if $c_m(u) \neq c_m(v)$ for every two distinct vertices u and v of G. That is, each such edge coloring of G induces a *multiset-irregular* vertex coloring of G. An edge coloring of G is a *kaleidoscopic coloring* if it is a k-kaleidoscopic coloring for some integer $k \geq 3$. Thus, a kaleidoscopic coloring is both *set-regular* and *multiset-irregular*. A regular graph G is called a k-*kaleidoscope* if G has a k-kaleidoscopic coloring. Figure 3.1 shows a 6-regular 3-kaleidoscope G of order 8 together with a 3-kaleidoscopic coloring of G, where the multiset-color of a vertex v is indicated inside the vertex v.

It is sometimes useful to look at kaleidoscopic colorings from another point of view. For a connected graph G of order $n \geq 3$ and a k-tuple factorization $\mathscr{F} = \{F_1, F_2, \cdots, F_k\}$ of G, where each F_i has no isolated vertices for $1 \leq i \leq k$, we associate the ordered k-tuple $a_1 a_2 \cdots a_k$ with a vertex v of G where $\deg_{F_i} v = a_i$ for $1 \leq i \leq k$. Thus $\sum_{i=1}^{k} \deg_{F_i} v = \deg_G v$. If distinct vertices have distinct k-tuples,

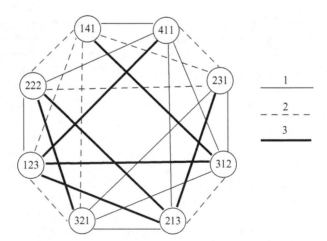

Fig. 3.1 A 6-regular 3-kaleidoscope G of order 8

then we can assign the color i ($1 \leq i \leq k$) to each edge of F_i and obtain a k-kaleidoscopic coloring of G for which the multiset-color $c_m(v)$ of v is $a_1 a_2 \cdots a_k$. In this case, the factorization $\mathscr{F}$ is called *irregular*. Conversely, every k-kaleidoscopic coloring of G gives rise to an irregular k-tuple factorization $\mathscr{F} = \{F_1, F_2, \cdots, F_k\}$ of G where the edges of F_i are those edges of G colored i and each F_i has no isolated vertices for $1 \leq i \leq k$. Hence, an edge coloring of a graph G is a kaleidoscopic coloring if and only if the corresponding factorization of G is irregular. Therefore, a graph G has a k-kaleidoscopic coloring if and only if G has an irregular k-tuple factorization.

3.2 Complete Kaleidoscopes

We begin with some observations. Let G be an r-regular k-kaleidoscope of order n. Then $k \leq r < n$. First, it is impossible that $r = k$, for otherwise, any edge coloring c of G in which every vertex of G is incident with at least one edge of each color results in $c_m(v)$ being the k-tuple in which each term is 1. If $r = k+1$, then there are at most k distinct k-tuples, each of which has 2 as one term and 1 for all other terms. In this case, $n \leq k$, which is impossible. Therefore, $r \geq k + 2$. Since the number of r-element multisets M whose elements belong to a k-element set S is $\binom{r-1}{r-k}$, we have the following bounds involving k, r and n.

Proposition 3.2.1. *If G is an r-regular k-kaleidoscope of order n, then*

$$k + 2 \leq r < n \leq \binom{r-1}{r-k} = \binom{r-1}{k-1}.$$

Proof. Let c be a k-kaleidoscopic coloring of G. Since we have already observed that $r \geq k+2$, it remains to show that $n \leq \binom{r-1}{r-k}$. The number of r-element multisets whose elements belong to the k-element set $[k]$ such that each multiset contains at least one element i for each i ($1 \leq i \leq k$) is

$$\binom{(r-k)+k-1}{r-k} = \binom{r-1}{r-k} = \binom{r-1}{k-1}.$$

Hence, $n \leq \binom{r-1}{k-1}$. □

As Proposition 3.2.1 indicates, for a given integer $k \geq 3$, the smallest possible value of r for an r-regular k-kaleidoscope is $r = k + 2$ and the smallest possible order n of such a graph is $n = r+1$. Obviously, the graph in question is the complete graph K_{k+3}. We show for each $k \geq 3$ that K_{k+3} is, in fact, a k-kaleidoscope.

Theorem 3.2.2. *For each integer $k \geq 3$, the complete graph K_{k+3} is a k-kaleidoscope.*

Proof. We consider two cases, according to whether k is odd or k is even.

Case 1.   *k is odd.* Then $k = 2\ell + 1$ for some positive integer ℓ. Thus, $k + 3 = 2\ell + 4$. It is known that $K_{2\ell+4}$ can be decomposed into $\ell + 1$ Hamiltonian cycles $H_1, H_2, \ldots, H_{\ell+1}$ and a 1-factor F. For each i with $1 \le i \le \ell$, let there be given a proper coloring H_i with the two colors $2i - 1$ and $2i$. Furthermore, we assign the color $2\ell + 1$ to each edge of F. Currently, each vertex of $K_{2\ell+4}$ is incident with exactly one edge of each of the colors $1, 2, \ldots, 2\ell + 1 = k$ and incident with exactly two edges in $H_{\ell+1}$ that have not yet been assigned any color. Let

$$H_{\ell+1} = C = (v_1, v_2 \ldots, v_{2\ell+4}, v_{2\ell+5} = v_1).$$

For $1 \le i \le \ell + 2$, assign the color i to the two edges of C incident with v_{2i}. This completes the edge coloring c of $K_{2\ell+4}$ and results in $c_s(v_i) = [k]$ for $1 \le i \le k + 3$ and $c_m(v_i)$ equalling the multiset M_i, where

- M_1 contains two elements 1, two elements $\ell + 2$ and one element of $[k] - \{1, \ell + 2\}$;
- M_{2i+1} contains two elements i, two elements $i + 2$ and exactly one element of $[k] - \{i, i + 1\}$ for $1 \le i \le \ell + 1$;
- M_{2i} contains three elements i and one element of $[k] - \{i\}$ for $1 \le i \le \ell + 2$.

Thus, c is set-regular and multiset-irregular and so K_{k+3} is a k-kaleidoscope when k is odd.

Case 2.   *k is even.* Then $k = 2\ell + 2$ for some positive integer ℓ. Thus, $k + 3 = 2\ell + 5$. Let $G = K_{2\ell+5}$, let $v \in V(G)$ and let $G' = G - v = K_{2\ell+4}$. As in Case 1, the graph G' can be decomposed into $\ell + 1$ Hamiltonian cycles $H_1, H_2, \ldots, H_{\ell+1}$ and a 1-factor F. Color the edges of $H_1, H_2, \ldots, H_\ell$ and F as in Case 1. At this point, each vertex of G' is incident with exactly one edge of each of the colors $1, 2, \ldots, 2\ell + 1$ (and no edges colored $2\ell + 2$) and incident with two edges in $H_{\ell+1}$ that have not yet been assigned any color. Let

$$H_{\ell+1} = C = (v_1, v_2 \ldots, v_{2\ell+4}, v_1).$$

For $1 \le i \le \ell + 2$, assign the color i to the edge $v_{2i-1}v_{2i}$ and the color $2\ell + 2$ to all other edges of C. This completes the edge coloring c' of $G' = K_{2\ell+4}$ and results in $c'_s(v_j) = [k]$ for $1 \le j \le 2\ell + 4$ and $c'_m(v_j) = M'_j$ for $1 \le j \le 2\ell + 4$ where $M'_{2i-1} = M'_{2i}$ contains two elements i and one element of $[k] - \{i\}$. We now consider the graph G. The edge coloring $c : E(G) \to [k]$ is defined by

$$c(e) = \begin{cases} c'(e) & \text{if } e \in E(G') \\ i & \text{if } e = vv_{2i-1} \text{ for } 1 \le i \le \ell + 2 \\ \ell + 2 + i & \text{if } e = vv_{2i} \text{ for } 1 \le i \le \ell - 1 \\ 2\ell + 2 & \text{if } e = vv_{2i} \text{ for each } i \in \{\ell, \ell + 1, \ell + 2\}. \end{cases}$$

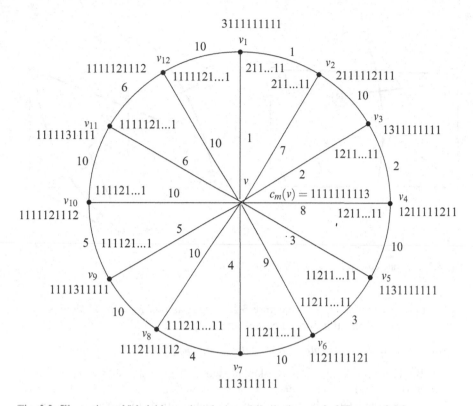

Fig. 3.2 Illustrating a 10-kaleidoscopic coloring of K_{13} in the proof of Theorem 3.2.2

This completes the edge coloring c of G and results in $c_s(x) = [k]$ for all $x \in V(G)$, $c_m(v_i) = M_i$ and $c_m(v) = M$, where

- M_{2i-1} is the only multiset containing three elements i for $1 \leq i \leq \ell + 2$,
- M_{2i} is the only multiset containing exactly two elements i for $1 \leq i \leq \ell + 2$ and
- M is the only multiset containing exactly one element i for each integer i with $1 \leq i \leq \ell + 2$.

This is illustrated in Fig. 3.2 for $k = 10$, where $c'_m(v_i) = M'_i$ is indicated inside the cycle C and $c_m(v_i) = M_i$ is indicated outside the cycle C for each i with $1 \leq i \leq 12$. Thus, c is set-regular and multiset-irregular and so $G = K_{k+3}$ is a k-kaleidoscope when k is even. $\qquad\square$

Not only is K_{k+3} a k-kaleidoscope but it is believed that every larger complete graph is also a k-kaleidoscope.

Conjecture 3.2.3. *For integers n and k with $n \geq k + 3 \geq 6$, the complete graph K_n is a k-kaleidoscope.*

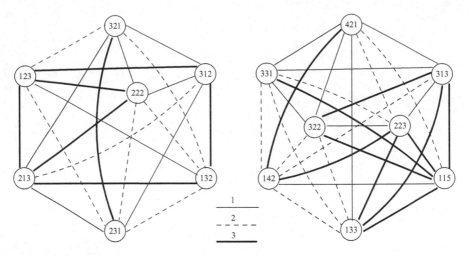

Fig. 3.3 A 3-kaleidoscopic coloring for each of K_7 and K_8

In the case where $k = 3$, Conjecture 3.2.3 suggests that K_n is a 3-kaleidoscope when $n \geq 6$. We verify this special case of the conjecture.

Theorem 3.2.4. *For each integer $n \geq 6$, the complete graph K_n is a 3-kaleidoscope.*

Proof. By Theorem 3.2.2, K_6 is a 3-kaleidoscope. Figure 3.3 shows that K_7 and K_8 are also 3-kaleidoscopes. Hence, we may assume that $n \geq 9$. Let $V(K_n) = \{v_1, v_2, \ldots, v_n\}$ and let F be the unique connected graph of order n containing exactly two vertices with equal degree. Without loss of generality, we may assume that

$$\deg_F v_i = \begin{cases} i & \text{if } 1 \leq i \leq \lfloor \frac{n}{2} \rfloor \\ i - 1 & \text{if } \lfloor \frac{n}{2} \rfloor + 1 \leq i \leq n. \end{cases}$$

Thus, $v_{\lfloor \frac{n}{2} \rfloor}$ and $v_{\lfloor \frac{n}{2} \rfloor + 1}$ are the only two vertices of F having the same degree $\lfloor \frac{n}{2} \rfloor$. To define an irregular factorization $\{F_1, F_2, F_3\}$ of K_n, we consider two cases, according to whether n is even or n is odd.

Case 1.   *n is even.* Then $n = 2p$ for some integer $p \geq 5$. Let

$$M = \{v_2 v_{n-1}, v_3 v_{n-2}, \ldots, v_p v_{p+1}\} = \{v_i v_{n-i+1} : 2 \leq i \leq p\}$$

be a matching of size $p - 1$ in F. Let

$$F_1 = F - \{v_3 v_n, v_4 v_n, v_p v_n\} - M$$
$$F_2 = \overline{F} + v_p v_n - \{v_1 v_3, v_1 v_4, v_1 v_p, v_3 v_4\}$$
$$F_3 = K_n - E(F_1) - E(F_2).$$

Hence,

$$E(F_3) = \{v_1v_3, v_1v_4, v_1v_p, v_3v_4\} \cup \{v_3v_n, v_4v_n\} \cup M.$$

In F_1, $\deg_{F_1} v_1 = \deg_{F_1} v_2 = \deg_{F_1} v_3 = 1$ and $\deg_{F_1} v_{n-2} = \deg_{F_1} v_n = n - 4$, while the remaining vertices have distinct degrees in F_1. Since

$$\deg_{F_2} v_1 = n - 5, \ \deg_{F_2} v_2 = n - 3, \ \deg_{F_2} v_3 = n - 6,$$

$$\deg_{F_3} v_{n-2} = 1 \text{ and } \deg_{F_3} v_n = 2,$$

it follows that $\{F_1, F_2, F_3\}$ is an irregular factorization of K_n. By assigning color i to each edge in F_i for $i = 1, 2, 3$, we obtain a 3-kaleidoscopic coloring c for K_n. Figure 3.4 illustrates such a factorization for K_{10}, where the bold edges in F and in $\overline{F}$ play a special role in the creation of F_1, F_2, F_3. With this coloring c, the multiset-colors of the vertices of K_{10} are

$$c_m(v_1) = 153, \ c_m(v_2) = 171, \ c_m(v_3) = 144, \ c_m(v_4) = 234, \ c_m(v_5) = 342$$

$$c_m(v_6) = 441, \ c_m(v_7) = 531, \ c_m(v_8) = 621, \ c_m(v_9) = 711, \ c_m(v_{10}) = 612.$$

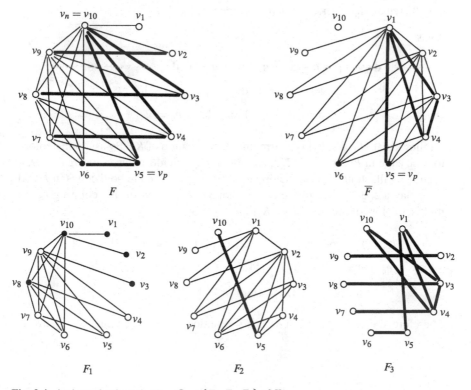

Fig. 3.4 An irregular factorization $\mathscr{F} = \{F_1, F_2, F_3\}$ of K_{10}

Case 2.   *n is odd.* Then $n = 2p + 1$ for some integer $p \geq 4$. Let

$$M = \{v_2 v_{n-1}, v_3 v_{n-2}, \ldots, v_p v_{p+2}\} = \{v_i v_{n-i+1} : 2 \leq i \leq p\}$$

be a matching of size $p - 1$ in F. Let

$$F_1 = F - \{v_3 v_n, v_4 v_n, v_{p+1} v_n\} - M$$

$$F_2 = \overline{F} + v_{p+1} v_n - \{v_1 v_3, v_1 v_4, v_1 v_{p+1}, v_2 v_{p+1}, v_3 v_4\}$$

$$F_3 = K_n - E(F_1) - E(F_2).$$

Hence,

$$E(F_3) = \{v_1 v_3, v_1 v_4, v_1 v_{p+1}, v_2 v_{p+1}, v_3 v_4\} \cup \{v_3 v_n, v_4 v_n\} \cup M.$$

Observe that

(1) $\deg_{F_1} v_1 = \deg_{F_1} v_2 = \deg_{F_1} v_3 = 1$,
(2) if $n = 9$ (or $p = 4$), then $\deg_{F_1} v_4 = 2$ and $\deg_{F_1} v_5 = 3$; while if $n \geq 11$
     (or $p \geq 5$), then $\deg_{F_1} v_p = \deg_{F_1} v_{p+1} = p - 1$,
(3) $\deg_{F_1} v_{n-2} = \deg_{F_1} v_n = n - 4$ and
(4) the remaining vertices have distinct degrees in F_1.

Since

$$\deg_{F_2} v_1 = n - 5, \ \deg_{F_2} v_2 = n - 4, \ \deg_{F_2} v_3 = n - 6,$$

$$\deg_{F_2} v_p = p, \ \deg_{F_2} v_{p+1} = p - 1,$$

$$\deg_{F_3} v_{n-2} = 1 \text{ and } \deg_{F_3} v_n = 2,$$

it follows that $\{F_1, F_2, F_3\}$ is an irregular factorization of K_n. By assigning color i
to each edge in F_i for $i = 1, 2, 3$, we obtain a 3-kaleidoscopic coloring c for K_n.
Figure 3.5 illustrates such a factorization for K_{11}, where the bold edges in F and
in $\overline{F}$ play a special role in the creation of F_1, F_2, F_3. With this coloring c, the
multiset-colors of the vertices of K_{11} are

$$c_m(v_1) = 163, \ c_m(v_2) = 172, \ c_m(v_3) = 154, \ c_m(v_4) = 244,$$

$$c_m(v_5) = 451 \ c_m(v_6) = 442, \ c_m(v_7) = 541, \ c_m(v_8) = 631,$$

$$c_m(v_9) = 721, \ c_m(v_{10}) = 811, \ c_m(v_{11}) = 712.$$

Therefore, the complete graph K_n is a 3-kaleidoscope for each integer $n \geq 6$.   □

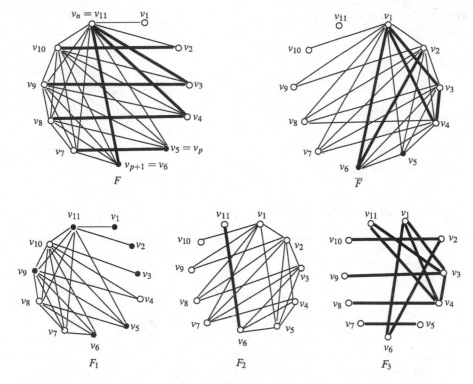

Fig. 3.5   An irregular factorization $\{F_1, F_2, F_3\}$ of K_{11}

3.3   3-Kaleidoscopes of Maximum Order

According to Proposition 3.2.1, the largest possible order of an r-regular 3-kaleidoscope is $\binom{r-1}{2}$. If $r \geq 7$ is an odd integer such that $r \equiv 3 \pmod 4$, then $\binom{r-1}{2}$ is odd and so no r-regular graphs of order $\binom{r-1}{2}$ exist for such odd integers r. On the other hand, there exists an r-regular 3-kaleidoscope of order $\binom{r-1}{2}$ for every integer $r \geq 5$ when $r \not\equiv 3 \pmod 4$.

Theorem 3.3.1. *For each integer $r \geq 5$ such that $r \not\equiv 3$ (mod 4), there exists an r-regular 3-kaleidoscope of order $\binom{r-1}{2}$.*

Proof. Let $r \geq 5$ be an integer with $r \not\equiv 3 \pmod 4$; so either r is even or $r \equiv 1 \pmod 4$. Since the proofs of these two situations are essentially the same, we only consider the case when $r \geq 6$ is even. We begin by constructing an r-regular graph G_r of order $\binom{r-1}{2}$.

For $1 \leq i \leq r - 2$, let V_i be a set of $r - 1 - i$ vertices and let

$$V(G_r) = \bigcup_{i=1}^{r-2} V_i.$$

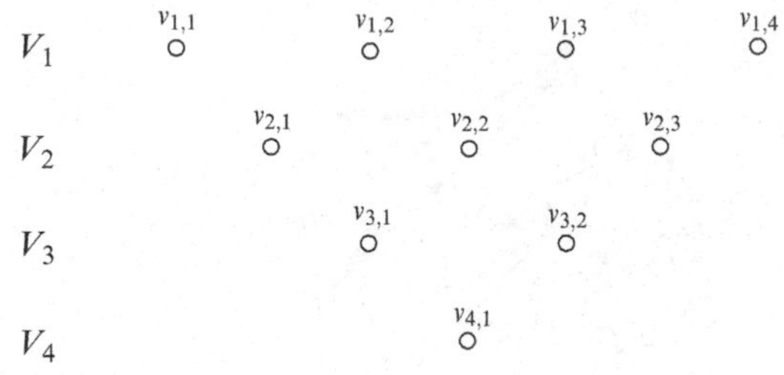

Fig. 3.6 The location of the vertices of the graph G_6

$$
\begin{array}{cccc}
 & v_{1,1} & v_{1,2} & v_{1,3} & v_{1,4} \\
V_1 & \circ & \circ & \circ & \circ \\
 & v_{2,1} & v_{2,2} & v_{2,3} & \\
V_2 & \circ & \circ & \circ & \\
 & v_{3,1} & v_{3,2} & & \\
V_3 & \circ & \circ & & \\
 & v_{4,1} & & & \\
V_4 & \circ & & &
\end{array}
$$

Thus, the order of G_r is

$$
|V(G_r)| = \sum_{i=1}^{r-2} |V_i| = \sum_{i=1}^{r-2} (r - i - 1) = \sum_{i=1}^{r-2} i = \binom{r-1}{2}.
$$

For $1 \leq i \leq r - 2$, let

$$
V_i = \{v_{i,j} : 1 \leq j \leq r - 1 - i\}.
$$

The vertices of G_r are placed in a triangular array such that for each i with $1 \leq i \leq r - 2$, the vertices of each set V_i are placed in a row where consecutive vertices are equally spaced two units apart, such that $V_1, V_2, \ldots, V_{r-2}$ are placed from top to bottom with each successive row of vertices one unit below the preceding. For $r = 6$, the vertices of $G_r = G_6$ are thus drawn as indicated in Fig. 3.6.

For $1 \leq i \leq r - 2$, we now construct a subgraph H_i of G_r with vertex set V_i. The graph H_{r-2} is the trivial graph; while for $1 \leq i \leq r - 3$, the graph H_i is the unique connected graph of order $r - 1 - i$ containing exactly two vertices of the same degree such that

$$
\deg_{H_i} v_{i,1} \leq \deg_{H_i} v_{i,2} \leq \cdots \leq \deg_{H_i} v_{i,r-1-i}.
$$

For $1 \leq i \leq r - 2$, let

$$
U_i = \{v_{i,1}, v_{i,2}, \ldots, v_{i,\lceil (r-2-i)/2 \rceil}\}
$$
$$
W_i = \{v_{i,\lceil (r-2-i)/2 \rceil}, v_{i,\lceil (r-2-i)/2 \rceil+1}, \ldots, v_{i,r-1-i}\}.
$$

Then $H_i[U_i]$ is empty and $H_i[W_i]$ is complete. We now add additional edges to obtain a subgraph of G_r, which we denote by F_2. Since $r \geq 6$ is even, it follows that $r = 2p + 2$ for some integer $p \geq 2$. Let A be an independent set of "slanted edges", defined by

$$
A = \begin{cases} \{v_{1,p+2i-1}\, v_{2,p+2i-2} : 1 \leq i \leq (p+1)/2\} & \text{if } p \text{ is odd} \\ \{v_{1,p+2i}\, v_{2,p+2i-1} : 1 \leq i \leq p/2\} & \text{if } p \text{ is even.} \end{cases}
$$

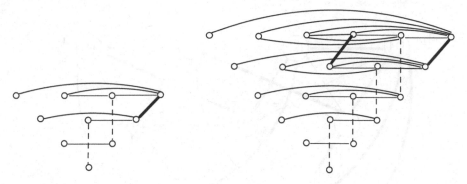

Fig. 3.7 The subgraph F_2 for $r = 6$ and $r = 8$

For each integer p and $1 \leq j \leq \lfloor p/2 \rfloor$, let E_j be an independent set of "vertical edges", defined by

$$E_j = \{v_{i,2p+2-i-2j} \, v_{i+2,2p+1-i-2j} : 1 \leq i \leq 2p + 2 - 4j\}.$$

We now add the edges in A and the edges in E_j ($1 \leq j \leq \lfloor p/2 \rfloor$) to those in the subgraphs H_i ($1 \leq i \leq r-2$) where all edges are straight line segments. All edges in each subgraph H_i are therefore "horizontal edges" for $1 \leq i \leq r-3$. This completes the construction of F_2. The graph F_2 is illustrated for both $r = 6$ and $r = 8$ in Fig. 3.7, where each slanted edge is indicated by a bold line and each vertical edge is indicated by a dashed line. Observe that $\deg_{F_2} v_{i,j} = j$ for all i and j.

The subgraph F_1 is obtained by rotating the subgraph F_2 clockwise through an angle of $2\pi/3$ radians; while the subgraph F_3 is obtained by rotating the subgraph F_2 counter-clockwise through an angle of $2\pi/3$ radians (or by rotating the subgraph F_2 clockwise through an angle of $4\pi/3$ radians). This completes the construction of G_r. This is illustrated in Fig. 3.8 for G_6, where each edge in F_1 is indicated by a thin solid line, each edge in F_2 is indicated by a dashed line and each edge in F_3 is indicated by a bold line. In Fig. 3.8, the label ijk inside the vertex $v_{i,j}$ indicates that

$$\deg_{F_1} v_{i,j} = i, \ \deg_{F_2} v_{i,j} = j \text{ and } \deg_{F_3} v_{i,j} = k = 6 - (i+j).$$

We now verify that in the rotation of F_2 into F_1 and F_3, every pair of adjacent vertices in F_2 is rotated into a pair of nonadjacent vertices of F_2 and, consequently, G_r is decomposed into F_1, F_2 and F_3. Let each vertex $v_{i,j}$ of G_r be denoted by $v_{i,j,k}$ as well, where $k = r - (i+j)$. Then the vertex set of G_r can be decomposed into $\lfloor (r-1)/3 \rfloor$ triangular sets X_ℓ ($1 \leq \ell \leq \lfloor (r-1)/3 \rfloor$) of vertices. For each such ℓ, the set X_ℓ consists of all those vertices $v_{i,j,k}$ such that $\ell \in \{i, j, k\}$. In Fig. 3.9, the sets X_1, X_2, X_3 are shown for G_{10}; while X_1 and X_2 are shown for G_8, where the vertices in X_1 are indicated by solid vertices, the vertices in X_2 are indicated by open vertices

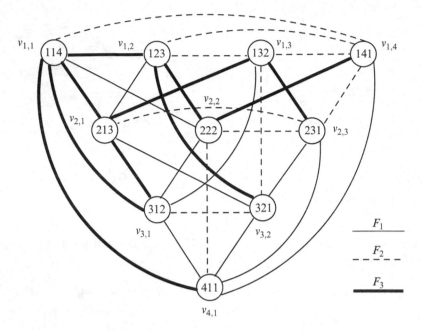

Fig. 3.8 The subgraphs F_1, F_2 and F_3 in G_6

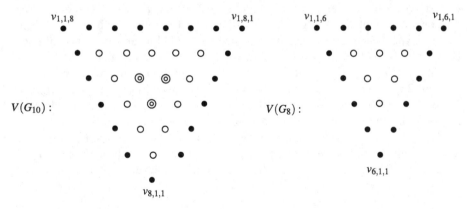

Fig. 3.9 The triangular sets X_ℓ for G_{10} and G_8

and the vertices in X_3 are indicated by double-circled vertices. The order of G_{10}, for example, is

$$\binom{9}{2} = 36 = \sum_{i=1}^{8} i = 1 + 2 + 3 + 4 + 5 + 6 + 7 + 8.$$

The number of vertices in X_1 is $8 + 7 + 6 = 21$, in X_2 is $5 + 4 + 3 = 12$ and in X_3 is $2 + 1 + 0 = 3$.

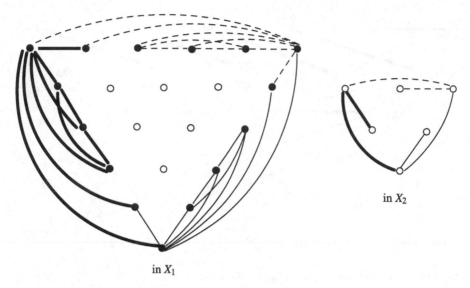

in X_1

Fig. 3.10 Edges in the two triangular sets X_1 and X_2 for G_8

The edges of G_r that join two vertices of X_ℓ ($1 \le \ell \le \lfloor (r-1)/3 \rfloor$) are those in H_ℓ (together with the edge $v_{1,r-2,1}v_{2,r-3,1}$ for $\ell = 1$), which is illustrated in Fig. 3.10. Thus, during the rotations of F_2 into F_1 and F_3, every two adjacent vertices of F_2 are rotated into nonadjacent vertices in F_1 or F_3. All other edges of F_2 join two different triangular sets.

The "vertical edges" in F_2 are rotated into edges of slopes -3 and 3; while the "slanted edges" are rotated into edges of slopes -1 and 0. This is illustrated in Fig. 3.11. For example, the vertical edge $v_{1,5}v_{3,4}$ in F_2 is rotated into the edge $v_{4,1}v_{5,2}$ in F_1 (of slope -3) and the edge $v_{1,3}v_{2,1}$ in F_3 (of slope 3). Furthermore, the slanted edge $v_{1,6}v_{2,5}$ in F_2 is rotated into the edge $v_{5,1}v_{6,1}$ in F_1 (of slope -1) and the edge $v_{1,1}v_{1,2}$ in F_3 (of slope 0). Hence, each edge of G_r belongs to exactly one of F_1, F_2, F_3.

By the construction of F_1, F_2 and F_3, it follows that

$$\deg_{F_t} v_{i,j,k} = \begin{cases} i & \text{if } t = 1 \\ j & \text{if } t = 2 \\ k & \text{if } t = 3. \end{cases}$$

Furthermore, if $\deg_{F_2} v_{i,j,k} = \deg_{F_2} v_{a,b,c}$ (and so $j = b$), then $\deg_{F_1} v_{i,j,k} \ne \deg_{F_1} v_{a,b,c}$ (that is, $i \ne a$). Hence, $\{F_1, F_2, F_3\}$ is an irregular factorization of G_r. Assigning the color t ($1 \le t \le 3$) to each edge of F_t, we obtain a 3-kaleidoscopic coloring of G_r for which the multiset-color $c_m(v_{i,j,k}) = ijk$ for all triples i, j, k. Therefore, G_r is an r-regular 3-kaleidoscope.   □

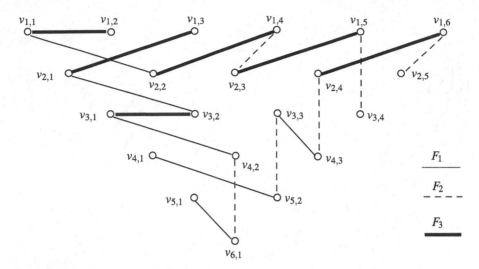

Fig. 3.11 Vertical and slanted edges in G_8

We close this chapter with the following conjecture.

Conjecture 3.3.2. *For each integer* $r \geq 7$ *where* $r \equiv 3$ (mod 4), *there is an* r-*regular* 3-*kaleidoscope of order* $\binom{r-1}{2} - 1$.

3.4   Majestic Edge Colorings

In 1985, Harary and Plantholt [44] introduced an unrestricted edge coloring c : $E(G) \rightarrow [k]$ of a graph G that induces a vertex-distinguishing coloring c', where $c'(v)$ is the set of colors of the edges incident with a vertex v. Such an edge coloring c is called a *set irregular edge coloring* of G. The minimum positive integer k for which a graph G has a set irregular edge coloring is the *set irregular chromatic index* of G and is denoted by si(G). (This parameter was referred to as the *point-distinguishing chromatic index* by Harary and Plantholt.) The set irregular chromatic index does not exist for K_2. Since every two vertices in a connected graph G of order $n \geq 3$ and size $m \geq 2$ are incident with different sets of edges, any edge coloring that assigns distinct colors of $[m]$ to the edges of G is a set irregular edge coloring. Hence, si(G) exists and si(G) $\leq m$.

In 2002, Zhang, Liu and Wang [77] introduced an edge coloring $c : E(G) \rightarrow [k]$ of a graph G for which both c and the induced set vertex coloring c' defined above are both proper. They referred to such a coloring c as an *adjacent strong edge coloring*. The minimum positive integer k for which G has an adjacent strong k-edge coloring is called the *adjacent strong chromatic index* of G.

Inspired by set irregular edge colorings and adjacent strong edge colorings, Gary Chartrand introduced unrestricted edge colorings $c : E(G) \to [k]$ of a graph G for which the induced vertex coloring c' defined above is proper. Such a coloring c is called a *majestic k-edge coloring*. The minimum positive integer k for which a graph G has a majestic k-edge coloring is called the *majestic chromatic index* of G. These concepts were studied in [10].

Typically, the graph coloring problems of greatest interest have been those of determining the minimum positive integer k for which it is possible to assign colors from the set $[k]$ to the vertices of a graph G in such a way that adjacent vertices are colored differently. For majestic edge colorings of a graph G, here too the goal is to determine the minimum positive integer k but, in this case, we are to assign colors from the set $[k]$ to the edges of G so that two adjacent vertices of G receive distinct induced colors. While the vertex colors are selected from the set $\mathscr{P}([k]) - \{\emptyset\}$ of nonempty subsets of $[k]$, it is of interest here as well to determine the minimum number of vertex colors satisfying these conditions. This leads to another concept. Suppose that G is a connected graph with majestic chromatic index $k \geq 2$. Then there exists a majestic k-edge coloring of G where the vertices of G are then colored with the nonempty subsets of $[k]$. Among all majestic k-edge colorings of G, the minimum number of nonempty subsets of $[k]$ needed to color the vertices of G so that two adjacent vertices of G are colored differently is called the *majestic chromatic number* of G. This concept was also studied in [10].

Chapter 4
Graceful Vertex Colorings

In Chap. 2 and 3, we described two edge colorings that give rise to two vertex colorings, one in terms of sets of colors and the other in terms of multisets. Now, in this chapter and the next, the situation is reversed, as we describe vertex colorings that give rise to edge colorings.

In a vertex labeling of a graph G, each vertex of G is assigned a label (an element of some set). If distinct vertices are assigned distinct labels, then the labeling is called *vertex-distinguishing* or *irregular*. That is, each vertex of G is uniquely determined by its label. A vertex labeling of G in which every two adjacent vertices are assigned distinct labels is a *neighbor-distinguishing labeling* or a *proper coloring*. Similarly, an edge labeling of G is *edge-distinguishing* if distinct edges are assigned distinct labels. An edge labeling of G in which every two adjacent edges are assigned distinct labels is an *edge-neighbor-distinguishing labeling* or a *proper edge coloring*. There are occasions when a vertex coloring (irregular or proper) of a graph gives rise to an edge-distinguishing labeling or a proper edge coloring, which is the subject of this chapter.

4.1 Graceful Labelings

In 1967, Alexander Rosa [67] introduced a vertex labeling of a graph that he called a *β-valuation*. In 1972, Solomon Golomb [41] referred to this labeling as a graceful labeling—terminology that has become standard. Let G be a graph of order n and size m. A *graceful labeling* of G is a one-to-one function $f : V(G) \rightarrow \{0, 1, \ldots, m\}$ that, in turn, assigns to each edge uv of G the label $f'(uv) = |f(u) - f(v)|$ such that no two edges of G are labeled the same. Therefore, if f is a graceful labeling of G, then the set of edge labels is $\{1, 2, \ldots, m\}$. A graph possessing a graceful labeling is a *graceful graph*. A major problem in this area is that of determining

P. Zhang, *A Kaleidoscopic View of Graph Colorings*, SpringerBriefs in Mathematics, DOI 10.1007/978-3-319-30518-9_4

which graphs are graceful. One of the best known conjectures dealing with graceful graphs involves trees and is due to Anton Kotzig (Rosa's doctoral advisor) and Gerhard Ringel (see [39]).

The Graceful Tree Conjecture *Every nontrivial tree is graceful.*

The *gracefulness* $\mathrm{grac}(G)$ of a graph G with $V(G) = \{v_1, v_2, \ldots, v_n\}$ is the smallest positive integer k for which it is possible to label the vertices of G with distinct elements of the set $\{0, 1, 2, \ldots, k\}$ in such a way that an edge is labeled as above and distinct edges receive distinct labels. The gracefulness of every such graph is defined, for if we label v_i by 2^{i-1} for $1 \le i \le n$, then a vertex labeling with this property exists. Thus, if G is a graph of order n and size m, then

$$m \le \mathrm{grac}(G) \le 2^{n-1}.$$

If $\mathrm{grac}(G) = m$, then G is graceful. The gracefulness of a graph G can therefore be considered as a measure of how close G is to being graceful—the closer the gracefulness is to m, the closer the graph is to being graceful. The exact values of $\mathrm{grac}(K_n)$ were determined for $1 \le n \le 10$ in [41]. For example, $\mathrm{grac}(K_4) = 6$, $\mathrm{grac}(K_5) = 11$ and $\mathrm{grac}(K_6) = 17$. The exact value of $\mathrm{grac}(K_n)$ is not known in general, however. On the other hand, Erdős showed that $\mathrm{grac}(K_n) \sim n^2$ (see [41]).

Graceful labelings have also been looked at in terms of colorings. A *rainbow vertex coloring* of a graph G of size m is an assignment f of distinct colors to the vertices of G. If the colors are chosen from the set $\{0, 1, \ldots, m\}$, resulting in each edge uv of G being colored $f'(uv) = |f(u) - f(v)|$ such that the colors assigned to the edges of G are also distinct, then this *rainbow* vertex coloring results in a *rainbow edge coloring* $f' : E(G) \to \{1, 2, \ldots, m\}$. So, such a rainbow vertex coloring is a graceful labeling of G.

The colorings of graphs that have received the most attention, however, are *proper vertex colorings* and *proper edge colorings*. In such a coloring of a graph G, every two adjacent vertices or every two adjacent edges are assigned distinct colors. As mentioned in Chap. 1, the minimum number of colors needed in a proper vertex coloring of G is its *chromatic number*, denoted by $\chi(G)$, while the minimum number of colors needed in a proper edge coloring of G is its *chromatic index*, denoted by $\chi'(G)$.

Inspired by graceful labelings, we now consider vertex colorings that induce edge colorings, both of which are proper rather than rainbow.

4.2   The Graceful Chromatic Number of a Graph

It is useful to describe notation for certain intervals of integers. For positive integers a, b with $a \le b$, let $[a, b] = \{a, a+1, \ldots, b\}$ and $[b] = [1, b]$. A *graceful k-coloring* of a nonempty graph G is a proper vertex coloring $c : V(G) \to [k]$, where $k \ge 2$, that induces a proper edge coloring $c' : E(G) \to [k-1]$ defined by $c'(uv) = |c(u) - c(v)|$.

Fig. 4.1 Graceful colorings
of K_4 and C_4

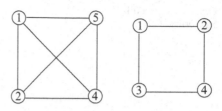

A vertex coloring c of a graph G is a *graceful coloring* if c is a graceful k-coloring
for some $k \in \mathbb{N}$. The minimum k for which G has a graceful k-coloring is called the
graceful chromatic number of G, denoted by $\chi_g(G)$. This concept was introduced
by Gary Chartrand and studied in [8, 9, 30]. There are immediate lower and upper
bounds for the graceful chromatic number of a graph.

Observation 4.2.1 ([8]). *If G is a nonempty graph of order n, then $\chi_g(G)$ exists
and*

$$\chi(G) \le \chi_g(G) \le \mathrm{grac}(G) \le 2^{n-1}.$$

Figure 4.1 shows two graceful graphs K_4 and C_4 together with a graceful coloring
for each of these two graphs. In fact, $\chi_g(K_4) = 5 < \mathrm{grac}(K_4) = 6$ and $\chi_g(C_4) = \mathrm{grac}(C_4) = 4$.

We make some additional useful observations. For a graceful k-coloring c of a
graph G, the *complementary coloring* $\bar{c} : V(G) \to [k]$ of G is a k-coloring defined
by $\bar{c}(v) = k + 1 - c(v)$ for each vertex v of G. If $xy \in E(G)$, then the color $\bar{c}'(xy)$
of xy induced by $\bar{c}$ is

$$\bar{c}'(xy) = |\bar{c}(x) - \bar{c}(y)| = |[(k+1) - c(x)] - [(k+1) - c(y)]|$$
$$= |c(x) - c(y)| = c'(xy).$$

This results in the following observation.

Observation 4.2.2 ([8]). *The complementary coloring of a graceful coloring of a
graph is also graceful.*

If c is a graceful k-coloring of a graph G, then the restriction of c to a subgraph
H of G is also a graceful coloring. Thus, we have the following observation.

Observation 4.2.3 ([8]). *If H is a subgraph of a graph G, then $\chi_g(H) \le \chi_g(G)$.*

If G is a disconnected graph having p components $G_1, G_2, \ldots, G_p$ for some
integer $p \ge 2$, then $\chi_g(G) = \max\{\chi_g(G_i) : 1 \le i \le p\}$. Thus, it suffices to consider
only nontrivial connected graphs.

For a vertex v in a graph G, let $N(v)$ be the *neighborhood* of v (the set of all
vertices adjacent to v in G) and let $N[v] = N(v) \cup \{v\}$ be the *closed neighborhood*
of v. If c is a graceful coloring of a nontrivial connected graph G and $v \in V(G)$,

Fig. 4.2 A graceful
5-coloring of Q_3

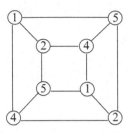

then c must assign distinct colors to the vertices in the closed neighborhood $N[v]$ of v. Thus, if $u, w \in V(G)$ such that $u \neq w$ and $d(u, w) \leq 2$, then $c(u) \neq c(w)$. Furthermore, if (x, y, z) is an $x - z$ path in G, where $c(x) > c(z)$, say, then $c(x) - c(y) \neq c(y) - c(z)$ and so $c(y) \neq \frac{c(x)+c(z)}{2}$. We state these useful observations next.

Observation 4.2.4 ([8]). *Let $c : V(G) \to [k]$, $k \geq 2$, be a coloring of a nontrivial connected graph G. Then c is a graceful coloring of G if and only if*

(i) for each vertex v of G, the vertices in the closed neighborhood $N[v]$ of v are assigned distinct colors by c and
(ii) for each path (x, y, z) of order 3 in G, $c(y) \neq \frac{c(x)+c(z)}{2}$.

As a consequence of condition (i) in Observation 4.2.4, it follows that if G is a nontrivial connected graph, then

$$\chi_g(G) \geq \Delta(G) + 1. \tag{4.1}$$

As an illustration, we determine $\chi_g(Q_3)$. Figure 4.2 shows a graceful 5-coloring of Q_3 and so $\chi_g(Q_3) \leq 5$. By (4.1), $\chi_g(Q_3) \geq 4$. Therefore, either $\chi_g(Q_3) = 4$ or $\chi_g(Q_3) = 5$. We show that $\chi_g(Q_3) \neq 4$. Assume, to the contrary, that Q_3 has a graceful 4-coloring using colors from the set [4]. By Observation 4.2.4, the four vertices in a 4-cycle in Q_3 must be colored differently. Thus, some vertex v of Q_3 is colored 3. However then, the three neighbors of v must be colored $1, 2, 4$, which implies that two incident edges of v are colored 1. This is impossible. Hence, $\chi_g(Q_3) = 5$.

This example also illustrates the following observation.

Observation 4.2.5. *If G is an r-regular graph where $r \geq 2$, then $\chi_g(G) \geq r + 2$.*

Since $\chi_g(K_{1,n-1}) = n = \Delta(K_{1,n-1}) + 1$, the bound in (4.1) is attained for all stars and, consequently, this bound is sharp. By Brooks' theorem [11],

$$\chi(G) \leq \Delta(G) + 1$$

for every graph G and, when G is connected, $\chi(G) = \Delta(G) + 1$ if and only if G is a complete graph or an odd cycle. Furthermore, by Vizing's theorem [74],

$$\chi'(G) \leq \Delta(G) + 1$$

for every nonempty graph G. Thus,

$$\chi_g(G) \geq \max\{\chi(G), \chi'(G)\}.$$

These observations together with Observation 4.2.5 yield the following.

Proposition 4.2.6. *If G is a nontrivial connected graph of order at least 3, then*

$$\chi_g(G) \geq \max\{\chi(G), \chi'(G)\} + 1.$$

Recall that the *distance* $d(u, v)$ between two vertices u and v in a connected graph G is the length of a shortest $u - v$ path in G; while the *diameter* diam(G) of a connected graph G is the largest distance between any two vertices of G. The following result is also a consequence of Observation 4.2.4.

Corollary 4.2.7 ([8]). *If G is a connected graph of order $n \geq 3$ with diameter at most 2, then $\chi_g(G) \geq n$.*

We saw that the star $K_{1,n-1}$, $n \geq 3$, is a graph of order n and diameter 2 having graceful chromatic number n. In fact, the star is one of many connected graphs having diameter 2 whose graceful chromatic number is its order.

Proposition 4.2.8 ([8]). *If G is a complete bipartite graph of order $n \geq 3$, then*

$$\chi_g(G) = n.$$

Proof. Let $G = K_{s,t}$ be a complete bipartite graph of order $n = s + t$ with partite sets U and W, where $U = \{u_1, u_2, \ldots, u_s\}$ and $W = \{w_1, w_2, \ldots, w_t\}$. Since the diameter of G is 2, it follows by Corollary 4.2.7 that $\chi_g(G) \geq n$. Next, consider a proper coloring $c : V(G) \to [n]$ defined by $c(u_i) = i$ for $1 \leq i \leq s$ and $c(w_j) = s+j$ for $1 \leq j \leq t$. Thus, $c'(u_iw_j) = |s + (j - i)|$ for $1 \leq i \leq s$ and $1 \leq j \leq t$. If i is fixed and $1 \leq j_1 \neq j_2 \leq t$, then $|s + (j_1 - i)| \neq |s + (j_2 - i)|$ and similarly, if j is fixed and $1 \leq i_1 \neq i_2 \leq s$, then $|s + (j - i_1)| \neq |s + (j - i_2)|$. Hence, c' is a proper edge coloring and c is a graceful n-coloring. Therefore, $\chi_g(G) = n$. □

In fact, there are also infinite classes of connected graphs G of order n such that diam$(G) = 2$ and $\chi_g(G) > n$.

Proposition 4.2.9 ([8]). *If G is a nontrivial connected graph of order n such that $\delta(G) > n/2$, then $\chi_g(G) > n$.*

Proof. Since $\delta(G) > n/2$, it follows that diam$(G) \leq 2$. Assume, to the contrary, that there is a graceful n-coloring c of G. By Observation 4.2.4, all vertices are assigned distinct colors by c and so there is a vertex v of G such that $c(v) = \left\lceil \frac{n}{2} \right\rceil$. Let $S = [1, \left\lceil \frac{n}{2} \right\rceil - 1]$ and $T = [\left\lceil \frac{n}{2} \right\rceil + 1, n]$, where then $|S| \leq |T| = n - \left\lceil \frac{n}{2} \right\rceil = \left\lfloor \frac{n}{2} \right\rfloor$. By Observation 4.2.4, at most one element in each set

$$\{\left\lceil\frac{n}{2}\right\rceil - i, \left\lceil\frac{n}{2}\right\rceil + i\} \text{ where } 1 \le i \le \left\lceil\frac{n}{2}\right\rceil - 1$$

can be used to color the vertices in $N(v)$. Hence, there are at most $\lfloor\frac{n}{2}\rfloor$ colors that are available for the vertices in $N(v)$. Since $\deg v > n/2 \ge \lfloor\frac{n}{2}\rfloor$, this is impossible. Therefore, $\chi_g(G) > n$.   □

4.3   Graceful Chromatic Numbers of Some Well-Known Graphs

First, we determine the graceful chromatic number of a cycle. It is useful to introduce some notation. Let $C_n = (v_1, v_2, \ldots, v_n, v_{n+1} = v_1)$ be a cycle of order $n \ge 3$ where $e_i = v_i v_{i+1}$ for $i = 1, 2, \ldots, n$. For a vertex coloring c of C_n, let

$$s_c = (c(v_1), c(v_2), \ldots, c(v_n)).$$

Similarly, for an edge coloring c' of C_n, let

$$s_{c'} = (c'(e_1), c'(e_2), \ldots, c'(e_n)).$$

Proposition 4.3.1 ([8]). *For each integer $n \ge 4$,*

$$\chi_g(C_n) = \begin{cases} 4 & \text{if } n \ne 5 \\ 5 & \text{if } n = 5. \end{cases}$$

Proof. Let $C_n = (v_1, v_2, \ldots, v_n, v_{n+1} = v_1)$ be a cycle of order $n \ge 4$ where $e_i = v_i v_{i+1}$ for $i = 1, 2, \ldots, n$. First, suppose that $n = 5$. Since $\text{diam}(C_5) = 2$, it follows by Corollary 4.2.7 that $\chi_g(C_5) \ge 5$. Define a vertex coloring c such that $s_c = (1, 5, 3, 4, 2)$. Then the induced edge coloring c' satisfies $s_{c'} = (4, 2, 1, 2, 1)$. Thus, c is a graceful 5-coloring and so $\chi_g(C_n) = 5$.

Next, suppose that $n \ne 5$. First, we show that $\chi_g(C_n) \ge 4$. Assume, to the contrary, that there is a graceful 3-coloring c of C_n, say $c(v_1) = 1$. Since c is a graceful coloring, $\{c(v_2), c(v_n)\} = \{2, 3\}$, say $c(v_2) = 2$ and $c(v_n) = 3$. However then, $c(v_3) = 3$ and so $c'(v_1 v_2) = c'(v_2 v_3) = 1$, which is impossible. Hence, $\chi_g(C_n) \ge 4$. It remains to define a graceful 4-coloring c of C_n.

- $n \equiv 0 \pmod 4$. For $n = 4$, let $s_c = (1, 2, 4, 3)$. Then $s_{c'} = (1, 2, 1, 2)$.
  For $n \ge 8$, let $s_c = (1, 2, 4, 3, \ldots, 1, 2, 4, 3)$. Then $s_{c'} = (1, 2, \ldots, 1, 2)$.
- $n \equiv 1 \pmod 4$. For $n = 9$, let $s_c = (1, 2, 4, 1, 2, 4, 1, 2, 4)$.
  So $s_{c'} = (1, 2, 3, 1, 2, 3, 1, 2, 3)$.
  For $n \ge 13$, let $s_c = (1, 2, 4, 3, \ldots, 1, 2, 4, 3, 1, 2, 4, 1, 2, 4, 1, 2, 4)$.
  Then $s_{c'} = (1, 2, 1, 2, \ldots, 1, 2, 1, 2, 3, 1, 2, 3, 1, 2, 3)$.

- $n \equiv 2 \pmod 4$. For $n = 6$, let $s_c = (1, 2, 4, 1, 2, 4)$. Then $s_{c'} = (1, 2, 3, 1, 2, 3)$. For $n \geq 10$, let $s_c = (1, 2, 4, 3, \ldots, 1, 2, 4, 3, 1, 2, 4, 1, 2, 4)$. Then $s_{c'} = (1, 2, 1, 2, \ldots, 1, 2, 1, 2, 3, 1, 2, 3)$.
- $n \equiv 3 \pmod 4$. In this case, $n \geq 7$. Let $s_c = (1, 2, 4, 3, \ldots, 1, 2, 4, 3, 1, 2, 4)$. Then $s_{c'} = (1, 2, 1, 2, \ldots, 1, 2, 1, 2, 3)$.

In each case, there is a graceful 4-coloring of C_n. Therefore, $\chi_g(C_n) = 4$ when $n \neq 5$.   □

It is easy to see that $\chi_g(P_4) = 3$. For $n \geq 5$, the following is a consequence of Proposition 4.3.1.

Proposition 4.3.2 ([8]). *For each integer $n \geq 5$, $\chi_g(P_n) = 4$.*

Proof. Let $P_n = (v_1, v_2, \ldots, v_n)$ where $n \geq 5$. For $n = 5$, a graceful 4-coloring c^* of P_5 is defined by $(c^*(v_1), c^*(v_2), c^*(v_3), c^*(v_4), c^*(v_5)) = (1, 2, 4, 1, 2)$ and so $\chi_g(P_5) \leq 4$. For $n \geq 6$, since P_n is a subgraph of C_n, it follows by Observation 4.2.3 and Proposition 4.3.1 that $\chi_g(P_n) \leq 4$. We show that $\chi_g(P_n) \neq 3$. Suppose that there is a graceful 3-coloring c of P_n. Necessarily, $c(v_3) \neq 2$ and so we may assume that $c(v_3) = 1$. Thus, $\{c(v_2), c(v_4)\} = \{2, 3\}$, say $c(v_2) = 2$. However then, $c(v_1) = 3$ and so $c'(v_1 v_2) = c'(v_2 v_3) = 1$, which is impossible. Therefore, $\chi_g(P_n) = 4$.   □

We now turn our attention to wheels W_n of order $n \geq 6$, constructed by joining a new vertex to every vertex of an $(n-1)$-cycle.

Theorem 4.3.3 ([8]). *If W_n is the wheel of order $n \geq 6$, then $\chi_g(W_n) = n$.*

Proof. Let $G = W_n$, where $C_{n-1} = (v_1, v_2, \ldots, v_{n-1}, v_1)$ and whose central vertex is v_0. By Corollary 4.2.7, $\chi_g(G) \geq n$. Thus, it suffices to show that G has a graceful n-coloring. Figure 4.3 shows a graceful n-coloring of W_n for $n = 6, 7, 8$, where the central vertex is colored 1 and the graceful n-coloring of W_n for $n = 7, 8$ is obtained from the graceful $(n-1)$-coloring of W_{n-1} by inserting a new vertex into the cycle C_{n-2} of W_{n-1}, joining this vertex to the central vertex and then assigning the color n to this vertex.

Next, we show that for a given graceful $(n-1)$-coloring of W_{n-1} for some integer $n \geq 7$, in which the central vertex is colored 1, there is an edge xy on

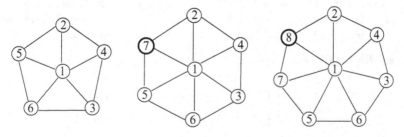

Fig. 4.3 Graceful colorings of W_6, W_7, W_8

the $(n-2)$-cycle C_{n-2} of W_{n-1} such that (1) a new vertex v can be inserted into the edge xy and joined to the central vertex v_0 of W_{n-1} to produce W_n and (2) the color n can be assigned to v to produce a graceful n-coloring of the resulting graph W_n. Now, let there be given a graceful $(n-1)$-coloring c of W_{n-1} for some integer $n \geq 7$, in which the central vertex is colored 1. It suffices to show that there exists an edge xy on C_{n-2} such that $c(x)$ and $c(y)$ satisfy the following two conditions:

(i) $c(x) \neq \frac{n+1}{2}$ and $c(y) \neq \frac{n+1}{2}$.

(ii) If (x', x, y, y') is a path on C_{n-2}, then $c(x) \neq \frac{c(x')+n}{2}$ and $c(y) \neq \frac{c(y')+n}{2}$.

Let $C_{n-2} = (v_1, v_2, \ldots, v_{n-2}, v_1)$. Since the diameter of W_{n-1} is 2, all vertices of W_{n-1} are assigned different colors by c. Hence, if $c(v_{i+1}) = \frac{c(v_{i+2})+n}{2}$ for some i, then $c(v_j) \neq \frac{c(v_{i+2})+n}{2}$ for all $j \neq i+1$ (where the subscripts are expressed as integers modulo $n-2$). We consider two cases.

Case 1. n *is odd.* Suppose that $c(v_{i+1}) = \frac{c(v_{i+2})+n}{2}$ for some i, in which case the edge $v_i v_{i+1}$ fails condition (ii). Since $n = 2c(v_{i+1}) - c(v_{i+2})$ is odd, it follows that $c(v_{i+2})$ is odd. Because there are $\frac{n-3}{2}$ vertices of C_{n-2} that are assigned odd colors by c (as the central vertex is colored 1), at most $\frac{n-3}{2}$ edges on C_{n-2} fail condition (ii). Hence, there are at least $(n-2) - \frac{n-3}{2} = \frac{n-1}{2} \geq 3$ edges on C_{n-2} that satisfy condition (ii). Among these edges that edges satisfy condition (ii), at most two of them fail condition (i). Thus, there is at least one edge xy on C_{n-2} such that $c(x)$ and $c(y)$ satisfy both (i) and (ii).

Case 2. n *is even.* Suppose that $c(v_{i+1}) = \frac{c(v_{i+2})+n}{2}$ for some i. Since $n = 2c(v_{i+1}) - c(v_{i+2})$ is even, it follows that $c(v_{i+2})$ is even. Because there are $\frac{n-2}{2}$ vertices on C_{n-2} that are assigned even colors by c, at most $\frac{n-2}{2}$ edges fail condition (ii). Hence, there are at least $(n-2) - \frac{n-2}{2} = \frac{n-2}{2} \geq 4$ edges that satisfy condition (ii). Since $(n+1)/2$ is not an integer, all of these edges satisfy condition (i) Therefore, there is at least one edge xy such that $c(x)$ and $c(y)$ satisfy both (i) and (ii).   □

For the regular complete bipartite graph $K_{p,p}$, it follows by Proposition 4.2.8 that $\chi_g(K_{p,p}) = 2p$. Since $\delta(K_{p,p}) = p = n/2$, the result stated in Proposition 4.2.9 is best possible. This suggests considering other regular complete multipartite graphs. For integers p and k where $p \geq 2$ and $k \geq 3$, let $K_{k(p)}$ be the regular complete k-partite graph, each of whose partite sets consists of p vertices. Thus, the order of $K_{k(p)}$ is $n = kp$ and the degree of regularity is $r = \frac{n(k-1)}{k} = (k-1)p$. The following result gives an upper bound for the graceful chromatic number of $K_{k(p)}$.

Theorem 4.3.4 ([8]). *For integers p and k where $p \geq 2$ and $k \geq 3$,*

$$\chi_g(K_{k(p)}) \leq \begin{cases} \left(2^{\frac{k+2}{2}} - 2\right)p - 2^{\frac{k-2}{2}} + 1 & \text{if } k \text{ is even} \\ \left(2^{\frac{k+3}{2}} - 3\right)p - 2^{\frac{k-1}{2}} + 1 & \text{if } k \text{ is odd.} \end{cases}$$

The upper bound for $\chi_g(K_{k(p)})$ presented in Theorem 4.3.4 is almost certainly not sharp. While $\chi_g(K_{p,p,p}) \leq 5p - 1$ for $p \geq 2$ according to Theorem 4.3.4, the following result gives an improved upper bound in this case. First, we introduce some useful notation. For a vertex coloring c of a graph G and a set X of vertices of G, let

$$c(X) = \{c(x) : x \in X\}$$

be the set of colors of the vertices of X.

Theorem 4.3.5 ([8]). *For each integer $p \geq 2$,*

$$\chi_g(K_{p,p,p}) \leq \begin{cases} 4p - 1 & \text{if } p \text{ is even} \\ 4p & \text{if } p \text{ is odd.} \end{cases}$$

Proof. Let $G = K_{p,p,p}$ with partite sets V_1, V_2, V_3, where $|V_i| = p$ for $1 \leq i \leq 3$. First, suppose that p is even. Define a proper coloring $c : V(G) \to [4p - 1]$ of G such that

$$c(V_1) = [p]$$
$$c(V_2) = \left[p + 1, 2p - \frac{p}{2}\right] \cup \left[2p + \frac{p}{2}, 3p - 1\right]$$
$$c(V_3) = [3p, 4p - 1].$$

To show that c is a graceful coloring of G, it suffices to show that if (x, z, y) is a path of order 3 in G, then

$$\frac{c(x) + c(y)}{2} \neq c(z). \tag{4.2}$$

Let $x \in V_i, y \in V_j, z \in V_t$, where $1 \leq i, j, t \leq 3$ and $t \neq i, j$. We may assume that $j \leq i$ and $c(y) \leq c(x)$.

- If $t < j$, then $c(z) < c(y) \leq \frac{c(x)+c(y)}{2}$.
- If $t > i$, then $\frac{c(x)+c(y)}{2} \leq c(x) < c(z)$.

Hence, we may assume that $j < t < i$ and so $j = 1, t = 2$ and $i = 3$. Observe that

$$\frac{c(x) + c(y)}{2} \geq \frac{3p + 1}{2} = 2p - \frac{p - 1}{2} > 2p - \frac{p}{2}$$
$$\frac{c(x) + c(y)}{2} \leq \frac{p + 4p - 1}{2} = 2p + \frac{p - 1}{2} < 2p + \frac{p}{2}.$$

Thus, (4.2) holds.

Next, suppose that p is odd. A proper coloring $c : V(G) \to [4p]$ of G is defined by $c(V_1) = [p], c(V_2) = \left[p + 1, 2p - \lceil \frac{p}{2} \rceil\right] \cup \left[2p + \lceil \frac{p}{2} \rceil, 3p\right]$ and $c(V_3) = [3p+1, 4p]$.

Let (x, z, y) be a path of order 3 in G. Suppose that $x \in V_i, y \in V_j, z \in V_t$, where $1 \leq i, j, t \leq 3$ and $t \neq i, j$. By an argument similar to the one used in Case 1, we may assume that $j = 1, t = 2$ and $i = 3$. Observe that

$$\frac{c(x) + c(y)}{2} \geq \frac{(3p + 1) + 1}{2} > \frac{3p + 1}{2} = 2p - \frac{p - 1}{2} > 2p - \left\lceil \frac{p}{2} \right\rceil$$

$$\frac{c(x) + c(y)}{2} \leq \frac{p + 4p}{2} < 2p + \frac{p + 1}{2} = 2p + \left\lceil \frac{p}{2} \right\rceil.$$

Thus, (4.2) holds.   □

Indeed, there is a reason to believe that the upper bound for $\chi_g(K_{p,p,p})$ presented in Theorem 4.3.5 is the actual value of $\chi_g(K_{p,p,p})$ for every integer $p \geq 2$.

Conjecture 4.3.6 ([8]). *For each integer $p \geq 2$,*

$$\chi_g(K_{p,p,p}) = \begin{cases} 4p - 1 & \text{if } p \text{ is even} \\ 4p & \text{if } p \text{ is odd.} \end{cases}$$

Conjecture 4.3.6 has been verified when $2 \leq p \leq 6$. As an illustration, we verify this for $p = 3$.

Proposition 4.3.7. $\chi_g(K_{3,3,3}) = 12$.

Proof. By Theorem 4.3.5, $\chi_g(K_{3,3,3}) \leq 12$. Hence, it remains to show that there is no graceful 11-coloring of $G = K_{3,3,3}$. Let V_1, V_2, V_3 be the partite sets of G. Assume, to the contrary, that G has a graceful coloring $c : V(G) \rightarrow [11]$. Since $\text{diam}(G) = 2$, no two vertices of G are assigned the same color. First, we claim that the color 6 cannot be used; for otherwise, say $6 \in c(V_1)$. Then at least one color in each of the five sets $\{i, 12 - i\}$ ($1 \leq i \leq 5$) is either not used by c or is in $c(V_1)$. Since $|c(V_1)| = 3$ and exactly two colors in [11] are not used by c, this is impossible. Thus, 6 is not used and so exactly nine of the ten colors in $[11] - \{6\}$ are used by c. We consider two cases.

Case 1. $5, 7 \in c(V(G))$. If $5, 7 \in c(V_i)$ for some $i = 1, 2, 3$, say $i = 1$, then one color in each of the four sets $\{1, 9\}, \{2, 8\}, \{3, 11\}, \{4, 10\}$ is either in $c(V_1)$ or is not used by c. Since $|c(V_1)| = 3$ and exactly one color in $[11] - \{6\}$ is not used c, this is impossible. Thus, we may assume that $5 \in c(V_1)$ and $7 \in c(V_2)$. Then the color 3 is either not used or is in $c(V_1)$ and the color 9 is either not used or is in $c(V_2)$. We may assume that $3 \in c(V_1)$ and so the color 4 is either not used or is in $c(V_1)$.

Subcase 1.1. $9 \in c(V_2)$. Then the color 8 is either not used or is in $c(V_2)$. We saw that the color 4 is either not used or is in $c(V_1)$. By symmetry, we may assume that $4 \in c(V_1)$. Then each of 10 and 11 is either not used or in $c(V_2)$. Therefore, each of the three colors 8, 10, 11 is either not used or is in $c(V_2)$. Since (i) $7, 9 \in c(V_2)$, (ii) at most one of 8, 10, 11 belongs to $c(V_2)$ and (iii) at most one of 8, 10, 11 is not used by c, at least one of 8, 10, 11 is in $c(V_3)$, a contradiction.

Subcase 1.2.   9 *is not used.* Then the colors used by c are $1, 2, 3, 4, 5, 7, 8, 10, 11$. Since $3, 4, 5 \in c(V_1)$, it follows that $c(V_1) = \{3, 4, 5\}$. Because $2, 8 \notin c(V_1)$, the vertex colored 5 is incident with two edges colored 3, a contradiction.

Case 2.   *Exactly one of* 5 *and* 7 *is used by* c, *say* 5. Then the colors used by c are $1, 2, 3, 4, 5, 8, 9, 10, 11$. We may assume that $5 \in c(V_1)$. Thus, at least one color in $\{2, 8\}$ and at least one color in $\{1, 9\}$ belongs to $c(V_1)$. Assume that $2 \in c(V_1)$. Thus, exactly one color in $\{1, 3\}$ belongs to $c(V_1)$. Since at least one color in $\{1, 9\}$ belongs to $c(V_1)$, it follows that $1 \in c(V_1)$ and so $c(V_1) = \{1, 2, 5\}$. However then, $3 \in c(V_2 \cup V_3)$ and the vertex colored 3 is incident with two edges colored 2, a contradiction. Thus, $2 \notin c(V_1)$ and so $8 \in c(V_1)$.

Next, suppose that $1 \in c(V_1)$. Thus, $c(V_1) = \{1, 5, 8\}$. However then, $3 \in c(V_2 \cup V_3)$ and the vertex colored 3 is incident with two edges colored 2, a contradiction. Thus, $1 \notin c(V_1)$ and so $9 \in c(V_1)$. Hence, $c(V_1) = \{5, 8, 9\}$. We may assume that $1 \in c(V_2)$. Since $5 \in c(V_1)$, it follows that $3 \in c(V_2)$ and so $2 \in c(V_2)$. Thus, $c(V_2) = \{1, 2, 3\}$ and $c(V_3) = \{4, 10, 11\}$. However then, the vertex colored 4 is incident with two edges colored 1, producing a contradiction. $\square$

The proof of Proposition 4.3.7 shows not only that $\chi_g(K_{3,3,3}) = 12$ but that there is a vertex coloring $c : V(G) \rightarrow [11]$ of $G = K_{3,3,3}$ that is a proper vertex coloring, namely $c(V_1) = \{5, 8, 9\}$, $c(V_2) = \{1, 2, 3\}$ and $c(V_3) = \{4, 10, 11\}$, whose induced edge coloring c' results only in one pair of adjacent edges having the same color.

4.4   The Graceful Chromatic Numbers of Trees

The graceful chromatic numbers of stars and paths were given in Propositions 4.2.8 and 4.3.2. We now discuss graceful colorings of other trees, beginning with the class of trees called caterpillars. A *caterpillar* is a tree T of order 3 or more, the removal of whose leaves produces a path (called the *spine* of T). Thus, every path, every star (of order at least 3) and every double star (a tree of diameter 3) is a caterpillar. The graceful chromatic numbers of caterpillars were determined in [8].

Theorem 4.4.1 ([8]).   *If T is a caterpillar with maximum degree $\Delta \geq 2$, then*

$$\Delta + 1 \leq \chi_g(T) \leq \Delta + 2.$$

Furthermore, $\chi_g(T) = \Delta + 2$ if and only if T has a vertex of degree Δ that is adjacent to two vertices of degree Δ in T.

It is a consequence of Theorem 4.4.1 then that there are trees T for which $\chi_g(T) = \Delta(T) + 1$ and trees T for which $\chi_g(T) = \Delta(T) + 2$. This brings up the question of whether there exists a tree T such that $\chi_g(T) - \Delta(T) > 2$. To answer

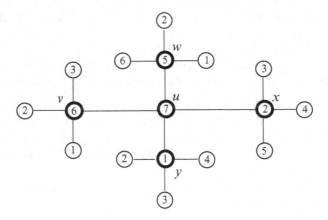

Fig. 4.4 A tree T_0 with $\chi_g(T_0) = \Delta(T_0) + 3$

this question, we consider the tree T_0 with $\Delta(T_0) = 4$ shown in Fig. 4.4. First, we claim that there is no graceful 6-coloring of T_0. Suppose that there is such a coloring $c : V(T_0) \to [6]$. The vertices in $N[u]$ are then colored with five colors from the set [6]. If $c(u) = 3$, then no two vertices in $N(u)$ can be colored both 2 and 4 or both 1 or 5 by Observation 4.2.4. Similarly, it is impossible that $c(u) = 4$. Thus, $c(u) \in \{1, 2, 5, 6\}$. The same can be said of v, w, x and y. This implies that two vertices of $N[u]$ are colored same, which is impossible. Since the 7-coloring of T_0 shown in Fig. 4.4 is a graceful coloring, it follows that $\chi_g(T_0) = 7 = \Delta(T_0) + 3$. For the tree T_0 in Fig. 4.4, observe that $\chi_g(T_0) = 7 = \left\lceil \frac{5\Delta(T_0)}{3} \right\rceil$. Indeed, for every tree T with maximum degree Δ, the graceful chromatic number of T can never exceed $\left\lceil \frac{5\Delta}{3} \right\rceil$, as we now show.

Theorem 4.4.2 ([8]). *If T is a nontrivial tree with maximum degree Δ, then*

$$\chi_g(T) \leq \left\lceil \frac{5\Delta}{3} \right\rceil.$$

Proof. Let $S_1 = \left[\left\lceil \frac{2\Delta}{3} \right\rceil\right]$, $S_2 = \left[\Delta + 1, \left\lceil \frac{5\Delta}{3} \right\rceil\right]$ and $S = S_1 \cup S_2$. In order to show that T has a graceful coloring using the colors in S, we first verify the following claim.

Claim. For each $a \in S$, there are at least Δ distinct elements $a_1, a_2, \ldots, a_\Delta \in S - \{a\}$ such that all of the Δ integers $|a - a_1|, |a - a_2|, \ldots, |a - a_\Delta|$ are distinct.

We consider three cases, according to the values of Δ modulo 3.

Case 1. $\Delta \equiv 0 \pmod 3$. Let $\Delta = 3k$ for some positive integer k. Then $\left\lceil \frac{2\Delta}{3} \right\rceil = 2k$ and so $S_1 = [2k]$ and $S_2 = [3k + 1, 5k]$. Let $a \in S$. By Observation 4.2.2, we may assume that $a \in S_1$. For each $i = 1, 2, \ldots, 2k$, let $a_i = 3k + i$. Then all of $|a - a_1|, |a - a_2|, \ldots, |a - a_{2k}|$ are distinct and $|a - a_i| = 3k + i - a \geq$

$3k + i - 2k = k + i \geq k + 1$ for $1 \leq i \leq 2k$. If $a \leq k$, then choose $a_{2k+j} = a + j$ for $1 \leq j \leq k$; while if $a \geq k + 1$, then choose $a_{2k+j} = a - j$ for $1 \leq j \leq k$. Then all of $|a - a_{2k+1}|, |a - a_{2k+2}|, \ldots, |a - a_{3k}|$ are distinct and $|a - a_{2k+j}| = j \leq k$. Since $|a - a_i| \geq k + 1$ for $1 \leq i \leq 2k$ and $|a - a_i| \leq k$ for $2k + 1 \leq i \leq 3k$, it follows that $|a - a_1|, |a - a_2|, \ldots, |a - a_{3k}|$ are distinct.

Case 2. $\Delta \equiv 1 \pmod 3$. Let $\Delta = 3k + 1$ for some nonnegative integer k. Then $\lceil \frac{2\Delta}{3} \rceil = 2k + 1$ and so $S_1 = [2k + 1]$ and $S_2 = [3k + 2, 5k + 2]$. Let $a \in S$. As observed in Case 1, we may assume that $a \in S_1$. For each $i = 1, 2 \ldots, 2k + 1$, let $a_i = 3k + 1 + i$. Then all of $|a - a_1|, |a - a_2|, \ldots, |a - a_{2k+1}|$ are distinct and $|a - a_i| = 3k + 1 + i - a \geq 3k + 1 + i - (2k + 1) = k + i \geq k + 1$ for $1 \leq i \leq 2k + 1$. If $a \leq k$, then choose $a_{2k+1+j} = a + j$ for $1 \leq j \leq k$; while if $a \geq k + 1$, then choose $a_{2k+1+j} = a - j$ for $1 \leq j \leq k$. Then all of $|a - a_{2k+2}|, |a - a_{2k+3}|, \ldots, |a - a_{3k+1}|$ are distinct and $|a - a_{2k+1+j}| = j \leq k$. Since $|a - a_i| \geq k + 1$ for $1 \leq i \leq 2k + 1$ and $|a - a_i| \leq k$ for $2k + 2 \leq i \leq 3k + 1$, it follows that $|a - a_1|, |a - a_2|, \ldots, |a - a_{3k+1}|$ are distinct.

Case 3. $\Delta \equiv 2 \pmod 3$. Let $\Delta = 3k + 2$ for some nonnegative integer k. Then $\lceil \frac{2\Delta}{3} \rceil = 2k + 2$ and so $S_1 = [2k + 2]$ and $S_2 = [3k + 2, 5k + 4]$. The argument is similar to the one in Case 2.

Therefore, the claim holds. It remains to construct a graceful coloring c of T using the colors in S. Let $v \in V(T)$ such that deg $v = \Delta$ and let

$$V_i = \{w \in V(T) : d(v, w) = i\} \text{ for } 0 \leq i \leq e(v),$$

where $e(v)$ is the eccentricity of v. Thus, $V_0 = \{v\}$ and $V_1 = N(v)$. Let $c(v) = a$ for some $a \in S$ and let $a_1, a_2, \ldots, a_\Delta \in S - \{a\}$ for which $|a - a_1|, |a - a_2|, \ldots, |a - a_\Delta|$ are distinct. Color the vertices of V_1 such that $\{c(w) : w \in V_1\} = \{a_1, a_2, \ldots, a_\Delta\}$. Thus, each vertex in $V_0 \cup V_1$ has been assigned a color from S such that all vertices and edges of the tree $T_1 = T[V_0 \cup V_1]$ are properly colored. Suppose then, for some integer i where $1 \leq i < e(v)$, that the colors of vertices in the tree

$$T_i = T\left[\cup_{j=0}^{i} V_j\right]$$

have been assigned colors from S such that all vertices and edges of T_i are properly colored. Next, we define the colors of vertices in V_{i+1}. Let $w \in V_i$ that is not an end-vertex of T. Suppose that deg $w = t \leq \Delta$ and $c(w) = b \in S$. Choose $b_1, b_2, \ldots, b_\Delta \in S - \{b\}$ such that $|b - b_1|, |b - b_2|, \ldots, |b - b_\Delta|$ are distinct. Let $u \in V_{i-1}$ such that $uw \in E(T)$. We may assume, without loss of generality, that

$$b_j \neq c(u) \text{ and } b_j \neq 2c(w) - c(u)$$

for $1 \leq j \leq t - 1 \leq \Delta - 1$. Color the vertices in $N(w) - \{u\} \subseteq V_{i+1}$ such that

$$\{c(w) : w \in N(w) - \{u\}\} = \{b_1, b_2, \ldots, b_{t-1}\}.$$

Continue this procedure for each non-end-vertex in V_i to define the color of each vertex in V_{i+1}. Therefore, T has a graceful coloring using colors from the set $S \subseteq \left[\left\lceil \frac{5\Delta}{3} \right\rceil\right]$ and so $\chi_g(T) \le \left\lceil \frac{5\Delta}{3} \right\rceil$.   □

We now describe a class of trees that will play a central role in our discussion. For each integer $\Delta \ge 2$, let $T_{\Delta,1}$ be the star $K_{1,\Delta}$. The *central vertex* of $T_{\Delta,1}$ is denoted by v. Thus, $\deg v = \Delta$ and all other vertices of $T_{\Delta,1}$ have degree 1. For each integer $h \ge 2$, let $T_{\Delta,h}$ be the tree obtained from $T_{\Delta,h-1}$ by identifying each end-vertex with the central vertex of the star $K_{1,\Delta-1}$. The tree $T_{\Delta,h}$ is therefore a rooted tree (with root v) having height h. The vertex v is then the *central vertex* of $T_{\Delta,h}$. In $T_{\Delta,h}$, every vertex at distance less than h from v has degree Δ; while all remaining vertices are leaves and are at distance h from v. Thus, $T_{2,2} = P_5$ and $T_{4,2}$ is the tree T_0 shown in Fig. 4.4. First, we determine the graceful chromatic number of $T_{\Delta,2}$ for each integer $\Delta \ge 2$.

Theorem 4.4.3 ([30]). *For each integer $\Delta \ge 2$,*

$$\chi_g(T_{\Delta,2}) = \left\lceil \frac{3\Delta + 1}{2} \right\rceil.$$

Proof. Let $T = T_{\Delta,2}$. Suppose that the central vertex of T is v and $N(v) = \{v_1, v_2, \ldots, v_\Delta\}$. For $i = 1, 2, \ldots, \Delta$, let $v_{i,1}, v_{i,2}, \ldots, v_{i,\Delta-1}$ be the $\Delta - 1$ end-vertices that are adjacent to v_i in T. We first show that $\chi_g(T_\Delta) \le \left\lceil \frac{3\Delta+1}{2} \right\rceil$. There are two cases, according to whether Δ is even or Δ is odd.

Case 1. Δ *is even.* Then $\Delta = 2k$ for some $k \in \mathbb{N}$ and so $\left\lceil \frac{3\Delta+1}{2} \right\rceil = 3k + 1$. Let $[3k + 1] = S_1 \cup S_2 \cup S_3$, where $S_1 = [k + 1]$, $S_2 = [k + 2, 2k]$ and $S_3 = [2k + 1, 3k + 1]$. Thus, $|S_1| = |S_3| = k + 1$ and $|S_2| = k - 1$. Define a coloring $c : V(T) \to [3k + 1]$ by

$$c(v) = k + 1, \ c(v_i) = i \text{ for } 1 \le i \le k \text{ and } c(v_i) = i + k + 1 \text{ for } k + 1 \le i \le 2k.$$

Hence, $\{c'(vv_i) : 1 \le i \le 2k\} = [2k]$. Next, for $1 \le i \le k$, let

$$\{c(v_{i,j}) : 1 \le j \le 2k - 1\} = [i + 1, i + 2k] - \{k + 1\};$$

while for $k + 1 \le i \le 2k$, let

$$\{c(v_{i,j}) : 1 \le j \le 2k - 1\} = [2k] - \{k + 1\}.$$

Thus, if $1 \le i \le k$, then $c'(vv_i) = k + 1 - i$ and $c'(v_i v_{i,j}) \ne k + 1 - i$; while if $k + 1 \le i \le 2k$, then $c'(vv_i) = i$ and $c'(v_i v_{i,j}) \ne i$. Therefore, c is a graceful coloring using the colors in $[3k + 1]$ and so $\chi_g(T) \le 3k + 1$.

Case 2. Δ *is odd.* Then $\Delta = 2k + 1$ for some $k \in \mathbb{N}$ and so $\left\lceil \frac{3\Delta+1}{2} \right\rceil = 3k + 2$. Let $[3k + 2] = S_1 \cup S_2 \cup S_3$, where $S_1 = [k + 1]$, $S_2 = [k + 2, 2k + 1]$ and $S_3 = [2k + 2, 3k + 2]$. Thus, $|S_1| = |S_3| = k + 1$ and $|S_2| = k$.

Define a coloring $c : V(T) \to [3k+2]$ by

$$c(v) = k+1, \; c(v_i) = i \text{ for } 1 \le i \le k \text{ and } c(v_i) = i+k+1 \text{ for } k+1 \le i \le 2k+1.$$

Hence, $\{c'(vv_i) : 1 \le i \le 2k+1\} = [2k+1]$. Next, for $1 \le i \le k$, let

$$\{c(v_{i,j}) : 1 \le j \le 2k\} = [i+1, i+2k+1] - \{k+1\};$$

while for $k+1 \le i \le 2k$, let

$$\{c(v_{i,j}) : 1 \le j \le 2k\} = [2k+1] - \{k+1\}.$$

Thus, if $1 \le i \le k$, then $c'(vv_i) = k+1-i$ $c'(v_i v_{i,j}) \ne k+1-i$; while if $k+1 \le i \le 2k+1$, then $c'(vv_i) = i$ and $c'(v_i v_{i,j}) \ne i$. Therefore, c is a graceful coloring using the colors in $[3k+2]$ and so $\chi_g(T) \le 3k+2$.

Next, we show that $\chi_g(T_\Delta) \ge \lceil \frac{3\Delta+1}{2} \rceil$. Again, we consider two cases, according to whether Δ is even or Δ is odd.

Case 1. Δ *is even.* Then $\Delta = 2k$ for some $k \in \mathbb{N}$ and so $\lceil \frac{3\Delta+1}{2} \rceil = 3k+1$. Assume, to the contrary, that there is graceful coloring c of T using colors from $[3k]$. Let $[3k] = S_1 \cup S_2 \cup S_3$, where $S_1 = [k]$, $S_2 = [k+1, 2k]$ and $S_3 = [2k+1, 3k]$. Thus, $|S_1| = |S_2| = |S_3| = k$. We claim that no vertex having degree $2k$ can be assigned a color in S_2; for otherwise, let $w \in N[v]$ such that $c(w) \in S_2$. Then $k+1 \le c(w) \le 2k$. Since $\deg w = 2k$, there is an edge incident with w, say wx, such that $|c(w) - c(x)| \ge 2k$. Hence, either $c(x) - c(w) \ge 2k$ or $c(w) - c(x) \ge 2k$. That is, either $c(x) \ge 2k + c(w) \ge 3k+1$ or $c(x) \le c(w) - 2k \le 0$, which is impossible. Therefore, every vertex in $N[v]$ must be assigned a color from $S_1 \cup S_3$. Since $|N[v]| = 2k+1$ and $|S_1 \cup S_3| = 2k$, a contradiction is produced.

Case 2. Δ *is odd.* Then $\Delta = 2k+1$ for some $k \in \mathbb{N}$ and so $\lceil \frac{3\Delta+1}{2} \rceil = 3k+2$. Assume, to the contrary, that there is graceful coloring c of T using colors from $[3k+1]$. Let $[3k+1] = S_1 \cup S_2 \cup S_3$, where $S_1 = [k]$, $S_2 = [k+1, 2k+1]$ and $S_3 = [2k+2, 3k+1]$. Thus, $|S_1| = |S_3| = k$ and $|S_2| = k+1$. We claim that no vertex having degree $2k+1$ can be assigned a color in S_2; for otherwise, let $w \in N[v]$ such that $c(w) \in S_2$. Then $k+1 \le c(w) \le 2k+1$. Since $\deg w = 2k+1$, there is an edge wx such that $|c(w) - c(x)| \ge 2k+1$. Hence, either $c(x) - c(w) \ge 2k+1$ or $c(w) - c(x) \ge 2k+1$. That is, either $c(x) \ge 2k+1+c(w) \ge 3k+2$ or $c(x) \le c(w)-2k-1 \le 0$, which is impossible. Therefore, every vertex in $N[v]$ must be assigned a color from $S_1 \cup S_3$. Since $|N[v]| = 2k+1$ and $|S_1 \cup S_3| = 2k$, a contradiction is produced. $\square$

Employing an approach similar to that used to verify Theorem 4.4.3, the following two results can be obtained.

Theorem 4.4.4 ([30]). *For each integer* $\Delta \ge 2$, $\chi_g(T_{\Delta,3}) = \lceil \frac{13\Delta+1}{8} \rceil$.

Theorem 4.4.5 ([30]). *For each integer* $\Delta \ge 2$, $\chi_g(T_{\Delta,4}) = \lceil \frac{53\Delta+1}{32} \rceil$.

The results obtained in Theorems 4.4.3–4.4.5 suggest the following conjecture.

Conjecture 4.4.6 ([30]). *For an integer $h \geq 2$, let $\sigma_h = 2^{2h-3} + \sum_{i=2}^{h} 2^{2i-4}$. Then*

$$\chi_g(T_{\Delta,h}) = \left\lceil \frac{\sigma_h \Delta + 1}{2^{2h-3}} \right\rceil.$$

In fact, for every integer $\Delta \geq 2$, the graceful chromatic number of the tree $T_{\Delta,h}$ is $\lceil \frac{5\Delta}{3} \rceil$ if its height h is sufficiently large.

Theorem 4.4.7 ([30]). *Let $\Delta \geq 2$ be an integer. If h is an integer with $h \geq 2 + \lfloor \frac{\Delta}{3} \rfloor$, then*

$$\chi_g(T_{\Delta,h}) = \left\lceil \frac{5\Delta}{3} \right\rceil.$$

The following two results are consequences of Theorem 4.4.7.

Corollary 4.4.8 ([30]). *For each integer $\Delta \geq 2$,*

$$\lim_{h \to \infty} \chi_g(T_{\Delta,h}) = \left\lceil \frac{5\Delta}{3} \right\rceil.$$

Corollary 4.4.9 ([30]). *If T is a tree with maximum degree $\Delta \geq 2$ containing a vertex v such that every vertex of T within distance $2 + \lfloor \frac{\Delta}{3} \rfloor$ of v also has degree Δ, then $\chi_g(T) = \lceil \frac{5\Delta}{3} \rceil$.*

With the aid of Theorem 4.4.7, we present a lower bound for the graceful chromatic number of a connected graph.

Corollary 4.4.10 ([30]). *If G is a connected graph with minimum degree $\delta \geq 2$, then*

$$\chi_g(G) \geq \left\lceil \frac{5\delta}{3} \right\rceil.$$

Proof. Assume, to the contrary, that there is a connected graph G with $\delta(G) = \delta \geq 2$ such that $\chi_g(G) \leq \lceil \frac{5\delta}{3} \rceil - 1$ and so G has a graceful ($\lceil \frac{5\delta}{3} \rceil - 1$)-coloring c : $V(G) \to [\lceil \frac{5\delta}{3} \rceil - 1]$. By Theorem 4.4.7, there exists a tree T with $\Delta(T) = \delta$ such that $\chi_g(T) = \lceil \frac{5\delta}{3} \rceil$. Let v be the central vertex (or root) of T. For $0 \leq i \leq e(v)$, let

$$V_i = \{x \in V(T) : d(v,x) = i\}.$$

Thus, $V_0 = \{v\}$ and $V_1 = N_T(v)$. Furthermore, let u be any vertex of G.

We now define a coloring c_T : $V(T) \to [\lceil \frac{5\delta}{3} \rceil - 1]$ of T from the graceful coloring c of G as follows. First, let $c_T(v) = c(u)$. Since c is a graceful coloring of G and $|N_T(v)| \le \Delta(T) = \delta = \delta(G) \le |c(N_G(u))|$, we can assign the colors from the set $c(N_G(u)) \subseteq \left[\left\lceil\frac{5\delta}{3}\right\rceil - 1\right]$ to the vertices in $N_T(v)$ such that the vertices and edges in the tree $T_1 = T[V_0 \cup V_1]$ are properly colored. Suppose then, for some integer i where $1 \le i < e(v)$, that the colors of vertices in the tree

$$T_i = T\left[\cup_{j=0}^i V_j\right]$$

have been assigned colors from $\left[\left\lceil\frac{5\delta}{3}\right\rceil - 1\right]$ such that

(i) for each $x \in V(T_i)$, there is $u_x \in V(G)$ for which $c_T(x) = c(u_x)$ and $c_T(N_{T_i}(x)) \subseteq c(N_G(u_x))$ and
(ii) all vertices and edges of T_i are properly colored.

Next, we define the colors of vertices in V_{i+1}. Let $y \in V_i$ that is not an end-vertex of T and let $z \in V_{i-1}$ such that $yz \in E(T)$. Then there is a vertex $u_y \in V(G)$ such that

$$c_T(y) = c(u_y) \text{ and } c_T(z) \in c_T(N_{T_i}(y)) \subseteq c(N_G(u_y)).$$

Since c is a graceful coloring and

$$|N_T(y) \cap V_{i+1}| \le \delta - 1 \le |c(N_G(u_y)) - \{c_T(z)\}|,$$

we can assign the colors from the set $c(N_G(u_y)) - \{c_T(z)\} \subseteq \left[\left\lceil\frac{5\delta}{3}\right\rceil - 1\right]$ to the vertices in $N_T(y) \cap V_{i+1}$ such that the vertices and edges of the tree $T_{i+1} = T\left[\cup_{j=0}^{i+1} V_j\right]$ are properly colored. Therefore, c_T is a graceful coloring of T using colors from the set $\left[\left\lceil\frac{5\delta}{3}\right\rceil - 1\right]$. However then, $\chi_g(T) \le \left\lceil\frac{5\delta}{3}\right\rceil - 1$, which is a contradiction.   □

The lower bound for the graceful chromatic number of a graph presented in Corollary 4.4.10 is best possible. For example, the graph G of Fig. 4.5 has $\delta(G) = \delta = 2$ and graceful chromatic number $\chi_g(G) = \left\lceil\frac{5\delta}{3}\right\rceil = 4$. A graceful 4-coloring of G is shown in the figure.

Fig. 4.5 A graph G with $\chi_g(G) = \left\lceil\frac{5\delta}{3}\right\rceil$

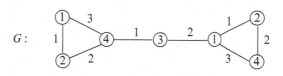

G:

Chapter 5
Harmonious Vertex Colorings

In this chapter, we study vertex colorings of graphs, where the colors are elements of $\mathbb{Z}_k$ or of $[k]$ for some integer $k \geq 2$. These give rise to either edge-distinguishing labelings or proper edge colorings defined in a variety of ways.

5.1 Harmonious Labelings

In 1980 (13 years after Rosa introduced 'graceful labelings'), Ronald Graham and Neil Sloane [42] introduced a vertex labeling of a graph they referred to as a harmonious labeling. For a connected graph G of size m, a *harmonious labeling* of G is an assignment f of distinct elements of the set $\mathbb{Z}_m$ of integers modulo m to the vertices of G so that the resulting edge labeling in which each edge uv of G is labeled $f(u) + f(v)$ (addition in $\mathbb{Z}_m$) is edge-distinguishing. Since such a vertex labeling is not possible if G is a tree, in this case, some element of $\mathbb{Z}_m$ is assigned to two vertices of G, while all other elements of $\mathbb{Z}_m$ are used exactly once. A graph that admits a harmonious labeling is called a *harmonious graph*.

The graphs H_1 and H_2, both of size 4, shown in Fig. 5.1 are harmonious. A harmonious labeling of each graph from the set $\mathbb{Z}_4$ is shown along with the resulting edge labels in that figure. The graph $H_3 = K_{2,3}$ of Fig. 5.1 is not harmonious, however. To see this, assume, to the contrary, that H_3 is harmonious. Since H_3 has size 6, there exists a harmonious labeling f of H_3 with elements of the set $\mathbb{Z}_6$. Suppose that $f(u_i) = a_i$ for $1 \leq i \leq 5$ (see Fig. 5.1). Thus, $\{a_1, a_2\}$ and $\{a_3, a_4, a_5\}$ are disjoint subsets of $\mathbb{Z}_6$. The edge labels of H_3 are therefore $a_i + a_j$, where $1 \leq i \leq 2$ and $3 \leq j \leq 5$. Since the edge labels of H_3 are distinct, $a_i + a_j = a_k + a_\ell$ (where $1 \leq i, k \leq 2$ and $3 \leq j, \ell \leq 5$) if and only if $i = k$ and $j = \ell$. This implies that $a_i - a_\ell = a_k - a_j$ if and only if $i = k$ and $j = \ell$. That is,

© The Author 2016
P. Zhang, *A Kaleidoscopic View of Graph Colorings*, SpringerBriefs in Mathematics,
DOI 10.1007/978-3-319-30518-9_5

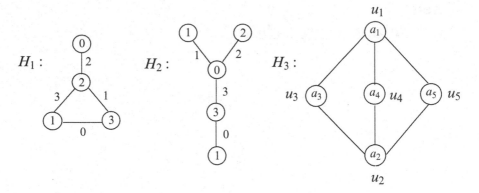

Fig. 5.1 Harmonious and non-harmonious graphs

$$\{a_i - a_j : \ 1 \le i \le 2 \text{ and } 3 \le j \le 5\} = \{0, 1, 2, 3, 4, 5\}.$$

In particular, for some i and j with $1 \le i \le 2$ and $3 \le j \le 5$, it follows that $a_i - a_j = 0$ and so $a_i = a_j$. This, however, is impossible since $\{a_1, a_2\}$ and $\{a_3, a_4, a_5\}$ are disjoint. Therefore, $H_3 = K_{2,3}$ is not harmonious. In fact, a similar argument provides the following.

Theorem 5.1.1. *For positive integers s and t with $s \le t$, the complete bipartite graph $K_{s,t}$ is harmonious if and only if $s = 1$.*

By Theorem 5.1.1, the only harmonious complete bipartite graphs are stars. Another large class of harmonious graphs is the odd cycles.

Theorem 5.1.2. *The cycle C_n is harmonious if and only if n is odd.*

Proof. Let $C_n = (v_0, v_1, \ldots, v_{n-1}, v_0)$ be a cycle of length n. Assume first that n is odd. Then $n = 2k + 1$ for some positive integer k. Consider the labeling that assigns v_i $(0 \le i \le n - 1)$ the label i. Then the k edges $v_i v_{i+1}$ $(0 \le i \le k - 1)$ are assigned all of the odd labels $1, 3, \ldots, n - 2$, while the $k + 1$ edges $v_i v_{i+1}$ $(k \le i \le n - 1)$ are assigned all of the even labels $0, 2, \ldots, n - 1$. Hence, C_n is harmonious.

Next, suppose, to the contrary, that C_n is harmonious for some even integer $n \ge 4$. Then $n = 2k$ for some integer $k \ge 2$ and there is a harmonious labeling of the cycle C_n, which assigns v_i the label a_i $(0 \le i \le n - 1)$. Thus,

$$\{a_0, a_1, \ldots, a_{n-1}\} = \{0, 1, \ldots, n - 1\}.$$

In $\mathbb{Z}_n$, let

$$s = \sum_{i=0}^{n-1} a_i = \sum_{i=0}^{n-1} i = \frac{n(n-1)}{2} = k(n - 1).$$

Hence,

$$\{a_0 + a_1, a_1 + a_2, \ldots, a_{n-1} + a_0\} = \{0, 1, \ldots, n - 1\}.$$

The sum in $\mathbb{Z}_n$ of the edge labels of C_n is therefore,

$$s = (a_0 + a_1) + (a_1 + a_2) + \cdots + (a_{n-1} + a_0) = 2 \sum_{i=0}^{n-1} i = 2s.$$

Thus, $2s \equiv s \pmod{n}$ and so $s \equiv 0 \pmod{n}$. Hence, $n \mid s$ and so $2k \mid k(n - 1)$. Thus $2 \mid (n - 1)$, which is impossible. $\qquad\square$

While it is easy to show that K_2, K_3, and K_4 are harmonious, Graham and Sloane verified that K_n is not harmonious when $n \geq 5$.

Theorem 5.1.3 ([42]). *A nontrivial complete graph K_n is harmonious if and only if* $2 \leq n \leq 4$.

While many classes of trees have been shown to be harmonious, it is not known whether all trees are harmonious. Graham and Sloane [42] made the following conjecture, which parallels that of a famous conjecture on graceful labelings of trees.

The Harmonious Tree Conjecture *Every nontrivial tree is harmonious.*

As we saw, in the case of a harmonious labeling of a graph, the vertex labeling f is vertex-distinguishing and the induced edge labeling f' is edge-distinguishing. It turns out that if we are to consider the situation when f is a proper coloring and require that f' is also proper, then it can be shown that the color set $\mathbb{Z}_k$ can be replaced by the set $[k] = \{1, 2, \ldots, k\}$ of k positive integers. That is, these two types of colorings, one of which uses the color set $\mathbb{Z}_k$ and the other uses the color set $[k]$, are the same colorings. In fact, this concept has been studied and referred to as a *star coloring* or a 2-*distance coloring* of a graph (see [33–35, 53], for example).

5.2 Harmonious Colorings

Let $c : V(G) \to [k]$ be a proper vertex coloring of a graph G. If *at most* one pair of adjacent vertices are colored i and j, where $i, j \in [k]$, then c is called a *harmonious coloring*. A harmonious coloring c of G therefore induces an edge labeling c' of G where the edge uv is assigned the label $c'(uv) = \{c(u), c(v)\}$, which is then a 2-element subset of the set of colors assigned to the vertices of G. The resulting edge coloring c' is edge-distinguishing. Since every coloring that assigns distinct colors to distinct vertices in a graph is a harmonious coloring, it follows that every graph has at least one harmonious coloring. The minimum positive integer k for which a graph G has a harmonious k-coloring is called the *harmonious chromatic number* of G and is denoted by $h(G)$.

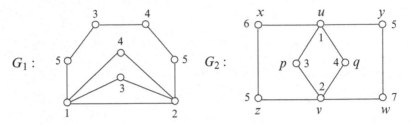

Fig. 5.2 Harmonious colorings of graphs

If G is a graph of size m with $h(G) = k$, then $m \leq \binom{k}{2} = k(k-1)/2$. Solving this inequality for k, we have $k \geq (1 + \sqrt{8m+1})/2$. This gives the following result.

Theorem 5.2.1. *If G is a graph of size m, then*

$$h(G) \geq \left\lceil \frac{1 + \sqrt{8m+1}}{2} \right\rceil.$$

According to Theorem 5.2.1, if G is a graph of size 10, then $h(G) \geq 5$. The two graphs G_1 and G_2 of Fig. 5.2 have size 10. While G_1 has harmonious chromatic number 5, the graph G_2 has harmonious chromatic number 7. The harmonious 5-coloring of G_1 in Fig. 5.2 shows that $h(G_1) = 5$, while the harmonious 7-coloring of G_2 in Fig. 5.2 shows only that $h(G_2) = 5$, $h(G_2) = 6$ or $h(G_2) = 7$. Suppose that there is a harmonious 5-coloring of G_2. Then the vertices u and v must be assigned distinct colors, say 1 and 2, respectively. Since there is only one pair of adjacent vertices colored 1 and i for $i = 2, 3, 4, 5$ and deg $u = 4$, only u can be colored 1. Similarly, only v can be colored 2. This, however, implies that two neighbors of u must be assigned the same color. This is impossible since the coloring is harmonious. Suppose next that there exists a harmonious 6-coloring of G_2. As before, we may assume that u and v are colored 1 and 2, respectively. Thus, p and q must be assigned distinct colors that are different from 1 and 2, say p and q are colored 3 and 4, respectively. Then x and y must be colored 5 and 6 as are z and w. Since the coloring is proper, the adjacent pairs $\{x, y\}$ and $\{w, y\}$ must both be colored 5 and 6. This, however, is impossible since the coloring is harmonious. Therefore, as claimed, $h(G_2) = 7$.

Since a given pair of distinct colors can be assigned to at most one pair of adjacent vertices in a harmonious coloring of a graph G, it follows that no two neighbors of a vertex in G can be assigned the same color. Hence, if v is a vertex for which deg $v = \Delta(G)$, then the neighbors of v must be assigned colors that are distinct from each other and from v. Consequently, we have the following.

Theorem 5.2.2. *For every graph G,*

$$h(G) \geq \Delta(G) + 1.$$

For the graphs G_1 and G_2 of Fig. 5.2, $\Delta(G_1) = \Delta(G_2) = 4$. Consequently, from Theorem 5.2.2, $h(G_1) \geq 5$ and $h(G_2) \geq 5$. Since both G_1 and G_2 have size $10 = \binom{5}{2}$, we have already observed these lower bounds. In fact, we have seen that $h(G_1) = 5$ and $h(G_2) = 7$.

For a graph G of order $n \geq 2$, $h(G) = 1$ if and only if $G = \overline{K}_n$. Furthermore, $h(K_n) = n$. However, there are noncomplete graphs of order n also having harmonious chromatic number n. Indeed, by Theorem 5.2.2, any graph of order n having maximum degree $n - 1$ has harmonious chromatic number n.

While we have seen some rather simple (although sharp) lower bounds for the harmonious chromatic number of a graph (in Theorems 5.2.1 and 5.2.2), a few more complex (although not sharp) upper bounds have been established as well. We describe some of these next.

A *partial harmonious coloring* of a graph G is a harmonious coloring of an induced subgraph of G such that no two neighbors of any uncolored vertex are assigned the same color. For a partial harmonious k-coloring of G, one of the k colors, say color i, is said to be *available* for an uncolored vertex v of G if v can be colored i and a new partial harmonious k-coloring of G results. For a color i to be available for v, no neighbor of v can be assigned the color i and no vertex of G can be colored i that is a neighbor of a vertex having the same color as a neighbor of v. The following result by Sin-Min Lee and John Mitchem [49] provides a lower bound for the number of available colors for an uncolored vertex in a partial harmonious coloring of a graph.

Theorem 5.2.3 ([49]). *If v is an uncolored vertex in a partial harmonious k-coloring of a graph G with $\Delta(G) = \Delta$ where each color class contains at most t vertices, then there are at least $k - t\Delta^2$ available colors for v.*

Proof. Assume first that every neighbor of v has been assigned a color. By hypothesis, no two neighbors of v are assigned the same color. As noted earlier, in order for one of the k colors to be unavailable for v, this color must either be (1) the color of a neighbor of v or (2) the color of a neighbor of a vertex having the same color as a neighbor of v. Let j be the color of some neighbor of v and let S_j be the color class consisting of those vertices of G that are colored j. Thus, $|S_j| \leq t$. Let $N(S_j)$ consist of all vertices of G that are neighbors of a vertex of S_j. Since the given coloring is a partial harmonious k-coloring of G, no color assigned to a vertex of $N(S_j)$ is available for v. Because $|N(S_j)| \leq t\Delta$, $v \in N(S_j)$ and v is uncolored, there are at most $t\Delta - 1$ unavailable colors for v of type (2) when considering $N(S_j)$.

Since there are at most Δ choices for a color i assigned to a neighbor of v, we see that there are at most $\Delta(t\Delta - 1) = t\Delta^2 - \Delta$ unavailable colors of type (2). However, the colors assigned to the neighbors of v are also unavailable for v. Hence, there are at most Δ unavailable colors of type (1). Thus, the total number of colors unavailable for v is at most $\Delta + (t\Delta^2 - \Delta) = t\Delta^2$. Therefore, the number of colors available for v is at least $k - t\Delta^2$.

If there are neighbors of v that are uncolored, then the argument above shows that the total number of colors available for v exceeds $k - t\Delta^2$ and so the result follows in both cases. □

With the aid of Theorem 5.2.3, an upper bound for the harmonious chromatic number of a graph was given by Lee and Mitchem in terms of the order and maximum degree of the graph.

Theorem 5.2.4 ([49]). *If G is a graph of order n having maximum degree Δ, then*

$$h(G) \leq \left(\Delta^2 + 1\right) \left\lceil \sqrt{n} \right\rceil.$$

Proof. If $\left(\Delta^2 + 1\right) \left\lceil \sqrt{n} \right\rceil \geq n$, then the result is obvious; so we may assume that

$$\left(\Delta^2 + 1\right) \left\lceil \sqrt{n} \right\rceil < n.$$

We claim that there is a harmonious coloring of G using $\left(\Delta^2 + 1\right) \left\lceil \sqrt{n} \right\rceil$ colors. Assume that this is not so. Then among all partial harmonious colorings of G, consider one where there is a harmonious coloring of an induced subgraph H of maximum order such that $\left(\Delta^2 + 1\right) \left\lceil \sqrt{n} \right\rceil$ colors are used in the coloring and each color class contains at most $\left\lceil \sqrt{n} \right\rceil$ vertices. Any coloring that assigns distinct colors to $\left(\Delta^2 + 1\right) \left\lceil \sqrt{n} \right\rceil$ vertices of G is a partial harmonious coloring of G, so partial harmonious colorings with the required properties exist. Now, because $H \neq G$, the graph G contains an uncolored vertex v.

By Theorem 5.2.3, v has at least

$$\left(\Delta^2 + 1\right) \left\lceil \sqrt{n} \right\rceil - \left\lceil \sqrt{n} \right\rceil \Delta^2 = \left\lceil \sqrt{n} \right\rceil$$

available colors. We claim that there exists an available color for v such that the color class consisting of the vertices assigned this color has fewer than $\left\lceil \sqrt{n} \right\rceil$ vertices. If this were not the case, then each of the $\left\lceil \sqrt{n} \right\rceil$ color classes consisting of the vertices assigned one of the available colors for v must contain $\left\lceil \sqrt{n} \right\rceil$ vertices. Since v belongs to none of these color classes, G must contain at least

$$\left\lceil \sqrt{n} \right\rceil \left\lceil \sqrt{n} \right\rceil + 1 \geq n + 1$$

vertices, which is impossible.

Thus, as claimed, there exists an available color i for v such that the color class consisting of the vertices colored i contains fewer than $\left\lceil \sqrt{n} \right\rceil$ vertices. By assigning v the color i, a partial harmonious coloring of G is produced, where there is a harmonious coloring of the induced subgraph $G[V(H) \cup \{v\}]$ whose order is larger than that of H and $\left(\Delta^2 + 1\right) \left\lceil \sqrt{n} \right\rceil$ colors are used in the coloring such that each color class contains at most $\left\lceil \sqrt{n} \right\rceil$ vertices. This contradicts the defining property of the given partial harmonious coloring. $\square$

Using partial harmonious colorings, Colin McDiarmid and Xinhua Luo [51] determined an improved upper bound for the harmonious chromatic number of a graph.

Theorem 5.2.5. *If G is a nonempty graph of order n $\geq$ 2 having maximum degree Δ, then*

$$h(G) \leq 2\Delta\sqrt{n-1}.$$

5.3 Harmonic Colorings

We have seen that a harmonious coloring of a graph G is a proper coloring of G having the property that if i and j are two distinct colors used in the coloring of G, then there is at most one pair of adjacent vertices assigned these two colors. A harmonious coloring c of G therefore induces an edge labeling of G where the edge uv is assigned the label $\{c(u), c(v)\}$, which is then a 2-element subset of the set of colors assigned to the vertices of G. Since no two edges of G are labeled the same, this vertex coloring is edge-distinguishing. That is, every harmonious coloring is edge-distinguishing.

There is a related edge-distinguishing vertex coloring of a graph in which adjacent vertices are permitted to be colored the same. While the early investigators of this concept referred to the coloring as a *line-distinguishing coloring* (see [37, 44], for example), this terminology doesn't "distinguish" it from a harmonious coloring. Consequently, the different but similar term, namely harmonic coloring, was used in [15]. A *harmonic coloring* of a graph G is a vertex coloring of G (where adjacent vertices may be assigned the same color) that induces the edge-distinguishing labeling that assigns to each edge uv the label $\{c(u), c(v)\}$, which is either a 2-element subset or a 1-element subset of colors, depending on whether $c(u) \neq c(v)$ or $c(u) = c(v)$. Since the coloring is edge-distinguishing, no two edges of G are labeled the same. The minimum positive integer k for which a graph G has a harmonic k-coloring is called the *harmonic chromatic number* or the *harmonic number* of G, which we denote by $h'(G)$. Thus,

$$h'(G) \leq h(G)$$

for every graph G. Furthermore, since no two neighbors of any vertex of G can be assigned the same color in a harmonic coloring of G, we have the following.

Theorem 5.3.1. *For every graph G,*

$$h'(G) \geq \Delta(G).$$

N. Zagaglia Salvi [69] showed that there are few graphs G for which $h'(G) = \Delta(G)$.

Theorem 5.3.2. *If G is a graph for which $h'(G) = \Delta(G)$ and v is a vertex of degree $\Delta(G)$, then at least one neighbor of v is an end-vertex.*

Proof. Since the result is true if $\Delta(G) = 1$, we may assume that $\Delta(G) = \Delta \geq 2$. Let u be a vertex with $\deg u = \Delta$. Then $|N[u]| = \Delta + 1 \geq 3$. Let a harmonic Δ-coloring of G be given. Then every two vertices of $N(u)$ are assigned distinct colors. Thus, u is assigned the same color as a vertex v adjacent to u.

We claim that $\deg v = 1$. Suppose that $\deg v \geq 2$. Then there is a vertex w distinct from u that is adjacent to v. Necessarily, $w \notin N(u)$ since uw and vw have distinct labels. Hence w is not adjacent to u and so w and a neighbor x of u are assigned the same color. This, however, implies that ux and vw are labeled the same, producing a contradiction. $\square$

If G is a graph of size m with $h'(G) = k$, then in a harmonic k-coloring of G, at most $\binom{k}{2}$ edges of G can be labeled with a 2-element set of distinct colors and at most k edges can be labeled with a 1-element set and so $m \leq k + \binom{k}{2} = \binom{k+1}{2}$. As a consequence of this observation, we have the following.

Theorem 5.3.3. *If G is a graph of size m, then*

$$h'(G) \geq \left\lceil \frac{-1 + \sqrt{1 + 8m}}{2} \right\rceil.$$

By Theorem 5.3.3 if a graph G has size $6 = \binom{3+1}{2}$, then $h'(G) \geq 3$. The two graphs G_1 and G_2 of Fig. 5.3 have six edges but $h'(G_1) = 3$ while $h'(G_2) = 4$. A harmonic 3-coloring of G_1 is shown in Fig. 5.3 together with a harmonic 4-coloring of G_2. To see why $h'(G_2) \neq 3$, first notice that any harmonic 3-coloring c must assign distinct colors to u and w (for otherwise uv and vw will be labeled the same). Suppose that $c(u) = 1$ and $c(w) = 2$. Then $c(v) \neq 1$ and $c(v) \neq 2$; so $c(v) = 3$. This implies that $c(x) = 1$ and $c(y) = 2$. However then, uw and xy are both labeled $\{1, 2\}$, which is impossible.

By Theorem 5.3.1, $h'(G) \geq \Delta(G)$ for every graph G. We have seen that the chromatic index $\chi'(G) \geq \Delta(G)$ as well. In fact, by Vizing's theorem,

$$\Delta(G) \leq \chi'(G) \leq \Delta(G) + 1.$$

Salvi [69] showed that $\chi'(G) = \Delta(G)$ whenever $h'(G) = \Delta(G)$.

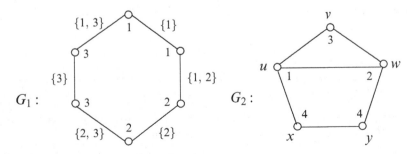

Fig. 5.3 Harmonious and non-harmonious graphs

Theorem 5.3.4. *If G is a graph with $h'(G) = \Delta(G)$, then*

$$\chi'(G) = \Delta(G).$$

Proof. If $h'(G) = \Delta(G) = 1$, then $G = K_2$ and $\chi'(G) = 1$. Hence, we may assume that $\Delta(G) \geq 2$. Suppose that $u_1, u_2, \ldots, u_p$ ($p \geq 1$) are the vertices of degree $\Delta(G)$ in G. By Theorem 5.3.2, each vertex u_i ($1 \leq i \leq p$) is adjacent to an end-vertex v_i. Let $H = G - \{v_1, v_2, \ldots, v_p\}$. Thus, $\Delta(H) = \Delta(G) - 1$. Since

$$\Delta(H) \leq \chi'(H) \leq \Delta(H) + 1$$

by Vizing's theorem, it follows that $\chi'(H)$ has one of two values. We consider these cases.

Case 1. $\chi'(H) = \Delta(H)$. Let a proper $\Delta(H)$-edge coloring of H be given. This produces an edge coloring of G except for the edges $u_i v_i$ ($1 \leq i \leq p$). By assigning a new color to each of these edges, a ($\Delta(H) + 1$)-edge coloring of G is obtained. Since $\Delta(H) + 1 = \Delta(G)$, it follows that $\chi'(G) = \Delta(G)$.

Case 2. $\chi'(H) = \Delta(H)+1$. Let a proper ($\Delta(H)+1$)-edge coloring of H be given. For each vertex u_i ($1 \leq i \leq p$), exactly one of the $\Delta(H)+1$ colors is not assigned to an edge incident with u_i. Assigning this color to $u_i v_i$ produces a ($\Delta(H) + 1$)-edge coloring of G. Since $\Delta(G) = \Delta(H) + 1$, it follows that $\chi(G) = \Delta(G)$. $\square$

As a consequence of Theorem 5.3.4, Salvi [69] showed that there is no graph G such that $\chi'(G) = \Delta(G) + 1$ and $h'(G) = \Delta(G)$.

Corollary 5.3.5. *For every graph G,*

$$h'(G) \geq \chi'(G).$$

Proof. Assume, to the contrary, that there exists a graph G such that $h'(G) < \chi'(G)$. Then $h'(G) = \Delta(G)$ and $\chi'(G) = \Delta(G) + 1$. This, however, contradicts Theorem 5.3.4. $\square$

While $\chi'(G) \leq \Delta(G)+1$ for every graph G, there are graphs G such that $h'(G) > \Delta(G) + 1$. For example, $h'(C_{10}) \geq 4$ by Theorem 5.3.3.

Chapter 6
A Map Coloring Problem

In Chaps. 2 and 3, we considered certain vertex colorings induced by edge colorings of a graph; while in Chaps. 4 and 5, we considered certain edge colorings induced by vertex colorings. Proper vertex colorings of graphs and the chromatic number of a graph originated from interest in map colorings. Other vertex colorings were also suggested from map colorings, which induced proper vertex colorings of graphs that often required using fewer colors than the chromatic number of a graph. In this chapter, we describe four such proper colorings originating from map colorings. These four vertex colorings will then be discussed in more detail in Chaps. 7–10.

6.1 A New Look at Map Colorings

The subject of graph colorings goes back to 1852 when the young British mathematician Francis Guthrie observed that the counties in a map of England could be colored with four colors so that every two adjacent counties are colored differently. This led to the Four Color Problem of determining whether the regions of *every* plane map could be colored with four or fewer colors in such a way that every two adjacent regions are colored differently. Of course, the Four Color Problem has an affirmative solution, as was announced in 1976 by Kenneth Appel and Wolfgang Haken. As a consequence of the resulting Four Color Theorem, it is possible to distinguish every two adjacent regions of every plane map M by coloring the regions of M with at most four colors.

For example, consider the map M of Fig. 6.1a. By the Four Color Theorem, there is a proper coloring of the regions of M with four colors, say 1, 2, 3, 4, that is, adjacent regions are colored differently. Such a coloring is shown in Fig. 6.1b. Therefore, every proper coloring of a map distinguishes every pair of adjacent regions. A different coloring of M is given in Fig. 6.1c, using the colors 1, 2, 3.

© The Author 2016
P. Zhang, *A Kaleidoscopic View of Graph Colorings*, SpringerBriefs in Mathematics,
DOI 10.1007/978-3-319-30518-9_6

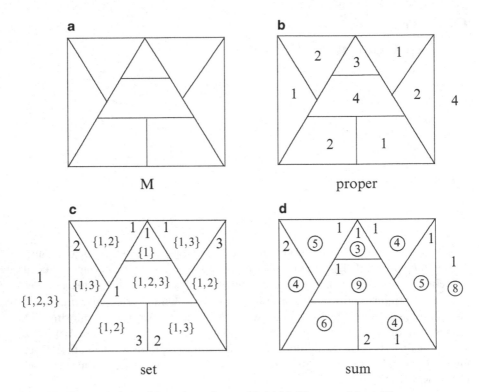

Fig. 6.1 Three colorings of the regions of a map M. (**a**) M, (**b**) proper, (**c**) set, (**d**) sum

Here too every two adjacent regions of M are distinguished from each other. In this case, the sets of colors of the neighboring regions of every two adjacent regions of M are different. A third coloring of M is given in Fig. 6.1d, using the colors 1 and 2. In this case as well, every two adjacent regions of M are distinguished from each other, where here the sums of the colors of the neighboring regions of every two adjacent regions of M are different.

Figure 6.1 therefore shows that it is possible to color the regions of a map M with fewer colors than that required of a proper coloring and still distinguish every two adjacent regions. As a second example, consider the map M shown in Fig. 6.2a. Here, as well, the regions of M can be properly colored with four colors, as shown in Fig. 6.2b, but with no fewer colors.

While there exists a 4-coloring of the regions of M shown in Fig. 6.2c so that the sets of colors of the neighboring regions of every two adjacent regions of M are different, there is no such 3-coloring and consequently there is no improvement in the number of colors needed for this map. On the other hand, there is a 3-coloring of the regions of M (using the colors 1, 2, 3), as shown in Fig. 6.2d, such that for every two adjacent regions of M, their distances to a nearest region of some color are not the same. Each region of M is labeled with a triple (a_1, a_2, a_3), where a_i

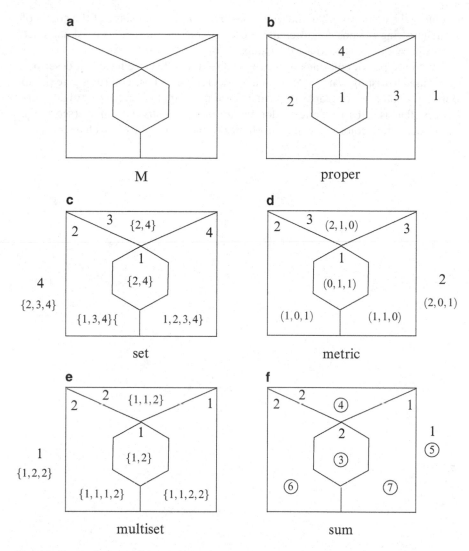

Fig. 6.2 Four colorings of the regions of a map M. (**a**) M, (**b**) proper, (**c**) set, (**d**) metric, (**e**) multiset, (**f**) sum

$(1 \leq i \leq 3)$ is the distance from that region to a nearest region colored i. Thus, adjacent regions can be distinguished by this coloring. There is no such 2-coloring that accomplishes this however. On the other hand, there does exist a 2-coloring of the regions of M with the colors 1 and 2, as shown in Fig. 6.2e, so that the multisets of the colors of the neighboring regions of every two adjacent regions of M are different. If the colors of the neighboring regions were summed, then we do not distinguish every two adjacent regions of M by this coloring. However, if we were to replace the color 1 by 2 in the centermost region, then the sums of the colors of the neighboring regions are different for every two adjacent regions of M and once again adjacent regions of M are distinguished by this coloring.

The four types of colorings of the regions of a map that we have just described can be used to distinguish every pair of adjacent regions. These colorings give rise to four vertex colorings of graphs that can be used to distinguish every pair of adjacent vertices, that is, all four vertex colorings are *neighbor-distinguishing* (see [25]). These four vertex colorings will be the topics of the four succeeding chapters.

Chapter 7
Set Colorings

If all of the vertices of a graph G of order n are distinguished as a result of a vertex coloring of G, then of course n colors are needed to accomplish this. On the other hand, if the goal of a graph coloring is only to distinguish every two adjacent vertices in G by means of a vertex coloring, then, of course, this can be accomplished by means of a proper coloring of G and the minimum number of colors needed to do this is the *chromatic number* $\chi(G)$ of G. Among the methods that can be used to distinguish every two adjacent vertices in G and that may require using fewer than $\chi(G)$ colors is the vertex coloring that assigns to each vertex the set of colors of its neighbors.

7.1 Set Chromatic Number

For a nontrivial connected graph G, let $c : V(G) \rightarrow \mathbb{N}$ be a vertex coloring of G where adjacent vertices may be assigned the same color. For a subset S of $V(G)$, define the set $c(S)$ of colors assigned to the vertices of S by

$$c(S) = \{c(v) : \ v \in S\}.$$

For a vertex v in a graph G, the *neighborhood color set* $c'(v) = c(N(v))$ of v is the set of colors of the neighbors of v. The coloring c is called a *set coloring* if $c'(u) \neq c'(v)$ for every pair u, v of adjacent vertices of G. Thus, a set coloring is a vertex coloring c that induces another vertex coloring c' of G defined by $c'(x) = c(N(x))$ for every vertex x of G. The minimum number of colors required of such a coloring c is called the *set chromatic number* of G and is denoted by $\chi_s(G)$. These concepts were introduced and studied in [20] and studied further in [40, 54].

For a graph G with chromatic number k, let c be a proper k-coloring of G. Suppose that u and v are adjacent vertices of G. Since $c(u) \in c'(v)$ and $c(u) \notin c'(u)$,

P. Zhang, *A Kaleidoscopic View of Graph Colorings*, SpringerBriefs in Mathematics, DOI 10.1007/978-3-319-30518-9_7

Fig. 7.1 A set coloring of a
graph

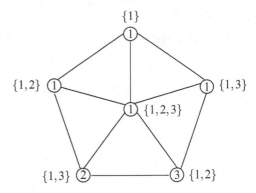

it follows that $c'(u) \neq c'(v)$. Hence, every proper k-coloring of G is also a set
k-coloring of G. Therefore, for every graph G,

$$\chi_s(G) \leq \chi(G). \tag{7.1}$$

Observe that if G is a connected graph of order n, then $\chi_s(G) = 1$ if and only if
$\chi(G) = 1$ (in this case $G = K_1$) and $\chi_s(G) = n$ if and only if $\chi(G) = n$ (in
this case $G = K_n$). Thus, if G is a nontrivial connected graph of order n that is not
complete, then

$$2 \leq \chi_s(G) \leq n - 1. \tag{7.2}$$

To illustrate these concepts, we consider the graph G (the wheel of order 6) in
Fig. 7.1. The chromatic number of G is $\chi(G) = 4$ and so $\chi_s(G) \leq 4$. In fact,
the set chromatic number of G is $\chi_s(G) = 3$, which we now verify. Figure 7.1
shows a set 3-coloring of G and so $\chi_s(G) \leq 3$. We now show that $\chi_s(G) \geq 3$.
Suppose, to the contrary, that there is a set 2-coloring c of G using the colors 1
and 2. Consider a triangle in G induced by three vertices v_1, v_2, v_3 of G. Since at
least two of these three vertices are colored the same, we may assume that two of
these vertices are assigned the color 1. Thus, $c'(v_i) = \{1\}$ or $c'(v_i) = \{1, 2\}$ for
each i $(1 \leq i \leq 3)$. This implies, however, that there are two adjacent vertices
having the same neighborhood color set, which contradicts our assumption that c is
a set coloring. Thus, $\chi_s(G) = 3$, as claimed.

The following observation will be useful to us.

Observation 7.1.1. *If u and v are two adjacent vertices in a graph G such that
$N(u) - \{v\} = N(v) - \{u\}$, then $c(u) \neq c(v)$ for every set coloring c of G.
Furthermore, if $S = N(u) - \{v\} = N(v) - \{u\}$, then $\{c(u), c(v)\} \nsubseteq c(S)$.*

7.2   The Set Chromatic Numbers of Some Classes of Graphs

Since every nonempty bipartite graph has chromatic number 2, the following is an immediate consequence of (7.1) and (7.2).

Observation 7.2.1. *If G is a nonempty bipartite graph, then $\chi_s(G) = 2$.*

In fact, if G is a nonempty graph, then $\chi_s(G) = 2$ if and only if G is bipartite, as we show next. We may restrict our attention to connected graphs.

Proposition 7.2.2. *If G is a connected graph with $\chi(G) \geq 3$, then $\chi_s(G) \geq 3$.*

Proof. Assume, to the contrary, that there exists a connected graph G with $\chi(G) \geq 3$ for which there exists a set 2-coloring $c : V(G) \to \{1, 2\}$. Since $\chi(G) \geq 3$, it follows that G contains an odd cycle $C = (v_1, v_2, \ldots, v_\ell, v_1)$, where $\ell \geq 3$ is an odd integer.

Consider the (cyclic) color sequence $s : c(v_1), c(v_2), \ldots, c(v_\ell), c(v_1)$. By a *block* of s, we mean a maximal subsequence of s consisting of terms of the same color. First, we claim that s cannot contain a block with an even number of terms; for suppose, without loss of generality, that $c(v_\ell) = 2$, $c(v_i) = 1$ for $1 \leq i \leq a$, where a is an even integer with $2 \leq a \leq \ell-1$, and $c(v_{a+1}) = 2$. Thus, $c'(v_i) \in \{\{1\}, \{1, 2\}\}$ for $1 \leq i \leq a$. Since $c'(v_1) = \{1, 2\}$ and c is a set coloring, it follows that

$$c'(v_i) = \begin{cases} \{1\} & \text{if } i \text{ is even} \\ \{1, 2\} & \text{if } i \text{ is odd} \end{cases}$$

for $1 \leq i \leq a$. However, this implies that $c'(v_a) = \{1\}$, which is impossible since $c(v_{a+1}) = 2$.

Hence, either (i) $c(v_i) = 1$ for all i ($1 \leq i \leq \ell$) or (ii) s contains an even number of blocks each of which has an odd number of terms. If (i) occurs, then $c'(v_i) \in \{\{1\}, \{1, 2\}\}$ for $1 \leq i \leq \ell$. Since ℓ is odd, there is an integer j ($1 \leq j \leq \ell$) such that $c'(v_j) = c'(v_{j+1})$, which is impossible. If (ii) occurs, then ℓ is even, which is also impossible.   □

The following three corollaries are immediate consequences of (7.1), Observation 7.2.1, and Proposition 7.2.2.

Corollary 7.2.3. *A nonempty graph G has set chromatic number 2 if and only if G is bipartite.*

Corollary 7.2.4. *If G is a 3-chromatic graph, then $\chi_s(G) = 3$.*

Corollary 7.2.5. *For each integer $n \geq 3$, $\chi_s(C_n) = \chi(C_n)$.*

We now determine the set chromatic number of every complete multipartite graph.

Proposition 7.2.6. *If G is a complete k-partite graph where $k \geq 2$, then $\chi_s(G) = k$.*

Proof. By (7.1), $\chi_s(G) \leq k$. Suppose that the statement is false. Then there is a smallest positive integer k for which there exists a complete k-partite graph G with

$\chi_s(G) \leq k - 1$. By Corollaries 7.2.3 and 7.2.4, $k \geq 4$. Suppose that the partite sets of G are $V_1, V_2, \ldots, V_k$. Let there be given a set $(k-1)$-coloring $c : V(G) \to [k-1]$ of G. We claim that for each partite set V_i $(1 \leq i \leq k)$ the coloring c_i, which is the restriction of c to the subgraph $G - V_i$, is a set coloring of $G - V_i$. In order to see that this is the case, let u and v be adjacent vertices in $G - V_i$. In G we have $c'(u) \neq c'(v)$. Since

$$c'(u) = c'_i(u) \cup c(V_i) \text{ and } c'(v) = c'_i(v) \cup c(V_i),$$

it follows that $c'_i(u) \neq c'_i(v)$. This implies that the coloring c_i of $G - V_i$ is a set coloring, as claimed. Since $\chi_s(G-V_i) = k-1$, it follows that $c(V(G)-V_i) = [k-1]$. Thus, $c'(x) = [k-1]$ for every vertex x of V_i. Since the partite set V_i was chosen arbitrarily, $c'(x) = [k-1]$ for every vertex x of G, which is impossible.   □

By Proposition 7.2.6, the complete k-partite graph $K_{1,1,\ldots,1,n-(k-1)}$ has set chromatic number k. This fact provides the following corollary.

Corollary 7.2.7. *For each pair k, n of integers with $2 \leq k \leq n$, there is a connected graph G of order n with $\chi_s(G) = k$.*

It is well known that the chromatic number of a graph G is at least as large as its clique number $\omega(G)$, which is the largest order of a clique (a complete subgraph) in G. The following observation will be useful to us.

Observation 7.2.8. *Let G be a graph of order $n \geq 2$.*

Then $\chi(G) = n - 1$ if and only if $\omega(G) = n - 1$.

Theorem 7.2.9. *For a connected graph G of order $n \geq 3$,*

$$\chi_s(G) = n - 1 \text{ if and only if } \chi(G) = n - 1.$$

Proof. If $\chi_s(G) = n - 1$, then $G \neq K_n$ by Proposition 7.2.6 and so the result follows immediately by (7.1). For the converse, assume that $\chi(G) = n - 1$. Then, by Observation 7.2.8, $\omega(G) = n - 1$ and so G is obtained from K_{n-1} by adding a new vertex u and joining u to some (but not all) vertices of K_{n-1}. Assume, to the contrary, that $\chi_s(G) = k \leq n - 2$ and let there be given a set k-coloring of G using the colors in $[k]$. Permuting the colors if necessary, we can obtain a set k-coloring $c : V(G) \to [k]$ such that $c(V(K_{n-1})) = [\ell]$, where $1 \leq \ell \leq k$. Since $\ell < n - 1$, some vertices in K_{n-1} are colored the same. Let $X \subseteq V(K_{n-1})$ such that for each $x \in X$, there exists a vertex $y \in X - \{x\}$ such that $c(y) = c(x)$. Hence, $|X| \geq 2$. Since each of the remaining $n - 1 - |X|$ vertices in K_{n-1} receives a unique color, it follows that $n - |X| \leq \ell$. For each $x \in X$, either (i) $c'(x) = [\ell]$ or (ii) $c'(x) = [\ell] \cup \{c(u)\}$ if $x \in N(u)$ and $c(u) \notin [\ell]$. This implies that $|X| \leq 2$. Hence, $|X| = 2$ and so $\ell = n - 2$. Then $k = \ell + 1$ (since $c(u) \notin [\ell]$) and $n - 2 = \ell = k - 1 \leq n - 3$, which is impossible.   □

For two graphs F and H, recall that the *union* of F and H is denoted by $F + H$, while the *join* of F and H is denoted by $F \vee H$. By Theorem 7.2.9 and its proof, a connected graph G of order $n \geq 3$ has $\chi_s(G) = n - 1$ if and only if $G = (K_{n-1-k} + K_1) \vee K_k$ for some integer k with $1 \leq k \leq n - 2$.

Corollary 7.2.10. *If G is a connected graph of order n such that*

$$\chi(G) \in \{1, 2, 3, n - 1, n\},$$

then $\chi_s(G) = \chi(G)$.

7.3   Lower Bounds for the Set Chromatic Number

We have already observed that $\chi_s(G) \leq \chi(G)$ for every graph G. There is also a lower bound for the set chromatic number of a graph in terms of its chromatic number.

Theorem 7.3.1. *For every graph G,*

$$\chi_s(G) \geq \lceil \log_2(\chi(G) + 1) \rceil.$$

Proof. Since this is true if $1 \leq \chi(G) \leq 3$, we may assume that $\chi(G) \geq 4$. Let $\chi_s(G) = k$ and let there be given a set k-coloring of G using the colors in $[k]$. Thus, $c'(x) \subseteq [k]$ for every vertex x of G. Since $c'(u) \neq c'(v)$ for every two adjacent vertices u and v of G, it follows that $c'(x)$ can be considered as a color for each $x \in V(G)$, that is, the coloring c of G defined by $c(x) = c'(x)$ for $x \in V(G)$ is a proper coloring of G. Since there are $2^k - 1$ nonempty subsets of $[k]$, it follows that c uses at most $2^k - 1$ colors. Therefore, $\chi(G) \leq 2^k - 1$ or $\chi(G) + 1 \leq 2^k$. Thus, $\chi_s(G) = k \geq \lceil \log_2(\chi(G) + 1) \rceil$, as desired. $\square$

By Corollary 7.2.10, the lower bound for the set chromatic number of a graph G in Theorem 7.3.1 is sharp if $\chi(G)$ is 1 or 2. If $\chi(G) = 3$, then $\chi_s(G) = 3 > \lceil \log_2(3 + 1) \rceil = 2$ and so this bound is not sharp in this case.

The Grötzsch graph G^* of Fig. 7.2 is known to have chromatic number 4. A set 3-coloring of G^* is also given in Fig. 7.2 and so $\chi_s(G^*) \leq 3$. By Proposition 7.2.2, $\chi_s(G^*) \geq 3$. Thus, $\chi_s(G^*) = 3$. Since $\lceil \log_2(\chi(G^*) + 1) \rceil = \lceil \log_2 5 \rceil = 3$, the lower bound for $\chi_s(G^*)$ is attained in this case.

While $\chi(G) \geq \omega(G)$ for every graph G, the clique number is not a lower bound for the set chromatic number of a graph.

Theorem 7.3.2. *For every graph G,*

$$\chi_s(G) \geq 1 + \lceil \log_2 \omega(G) \rceil. \tag{7.3}$$

Fig. 7.2 A set 3-coloring of
the Grötzsch graph

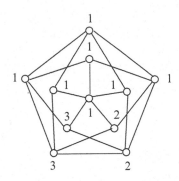

Proof. If $\omega(G) = 2$, then $\chi_s(G) \geq 2$; while if $\omega(G) = 3$, then $\chi_s(G) \geq 3$. Thus, we may assume that $\omega(G) = \omega \geq 4$. Let H be a clique of order ω in G with $V(H) = \{v_1, v_2, \ldots, v_\omega\}$. Suppose that $\chi_s(G) = k$ and let $c : V(G) \to [k]$ be a set k-coloring of G. We consider two cases, according to whether there are two vertices in $V(H)$ colored the same or no two vertices in $V(H)$ are assigned the same color.

Case 1. There are two vertices in $V(H)$ colored the same, say $c(v_1) = c(v_2) = 1$.
  Then $1 \in c'(v_i)$ for $1 \leq i \leq \omega$. Since there are exactly 2^{k-1} subsets of $[k]$ containing 1, it follows that $\omega \leq 2^{k-1}$ and so $k - 1 \geq \log_2 \omega$. Therefore, (7.3) holds.

Case 2. No two vertices in $V(H)$ are colored the same. Then ω distinct colors are used for the vertices in $V(H)$ and so $\omega \leq k$. Since $\omega \geq 4$, it follows that

$$k \geq \omega > 1 + \lceil \log_2 \omega(G) \rceil .$$

Again, (7.3) holds.   □

It was shown in [20] that the lower bound for the set chromatic number of a graph in Theorem 7.3.2 is sharp. Figure 7.3 shows a graph G with $\omega(G) = 4$ and $\chi_s(G) = 3$, and so $\chi_s(G) = 3 = 1 + \lceil \log_2 4 \rceil$.

While it is not known whether there is a graph G with $\chi_s(G) = a$ and $\chi(G) = b$ for *all* pairs a, b of integers with $2 \leq a \leq b$, it is known if $a \geq 1 + \log_2 b$. Should there exist a graph G with $\chi_s(G) = a$ and $\chi(G) = b$ where $a \geq 3$ and $a < 1 + \log_2 b$, then it follows by Theorem 7.3.2 that $\omega(G) < b$.

Theorem 7.3.3 ([40]). *For each pair a, b of integers with $2 \leq a \leq b \leq 2^{a-1}$, there exists a connected graph G with $\chi_s(G) = a$ and $\chi(G) = b$.*

For every connected graph G with $\chi_s(G) = a \geq 2$ that we have encountered, $\chi(G) \leq 2^{a-1}$. Thus, we present the following conjecture.

Conjecture 7.3.4 ([40]). *If G is a connected graph with $\chi_s(G) = a \geq 2$, then*

$$\chi(G) \leq 2^{a-1}.$$

Fig. 7.3 A graph G with
$\chi_s(G) = 1 + \lceil \log_2 \omega(G) \rceil$

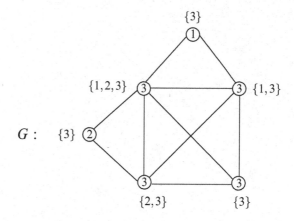

Chapter 8
Multiset Colorings

In the preceding chapter, a proper vertex coloring of a graph G was discussed that was defined from a given nonproper vertex coloring of G such that the color of a vertex is the set of colors of the neighbors of the vertex. In this chapter, proper vertex colorings are also discussed that arise from nonproper vertex colorings but here they are defined in terms of multisets rather than sets.

8.1 Multiset Chromatic Number

For a connected graph G and a positive integer k, let $c : V(G) \to [k]$ be a coloring of the vertices of G where adjacent vertices may be colored the same. The coloring c is called a *multiset coloring* if the multisets of colors of the neighbors of every two adjacent vertices of G are different, that is, for every two adjacent vertices u and v, there exists a color i such that the number of neighbors of u colored i and the number of neighbors of v colored i are not the same. For a vertex v of G, the multiset $M(v)$ of colors of the neighbors of v can be represented by a k-vector. The *multiset color code* of v is the k-vector

$$\text{code}_m(v) = (a_1, a_2, \cdots, a_k) = a_1 a_2 \cdots a_k,$$

where a_i is the number of occurrences of i in $M(v)$, that is, the number of vertices adjacent to v that are colored i for $1 \le i \le k$. Therefore,

$$\sum_{i=1}^{k} a_i = \deg v.$$

Thus, a vertex coloring of G is a multiset coloring if every two adjacent vertices have distinct multiset color codes. Hence, every multiset coloring of a graph G is a proper

© The Author 2016

P. Zhang, *A Kaleidoscopic View of Graph Colorings*, SpringerBriefs in Mathematics,
DOI 10.1007/978-3-319-30518-9_8

Fig. 8.1 A multiset
2-coloring of a 4-chromatic
graph G

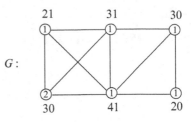

vertex coloring. The *multiset chromatic number* $\chi_m(G)$ of G is the minimum positive
integer k for which G has a multiset k-coloring. These concepts were introduced
and studied in [21]. For the 4-chromatic graph G of Fig. 8.1, the multiset chromatic
number of this graph is 2. Figure 8.1 shows a multiset 2-coloring of G together with
the multiset color code of each vertex of G.

Suppose that c is a proper vertex k-coloring of a graph G. If u is a vertex of G
and $c(u) = i$ for some integer i $(1 \leq i \leq k)$, then the i-th coordinate of the color
code of u is 0. On the other hand, if v is a neighbor of u, then the i-th coordinate of
the color code of v is at least 1, implying that $\text{code}_m(u) \neq \text{code}_m(v)$. Hence, every
proper coloring of G is a multiset coloring. Therefore, for every graph G of order n,

$$1 \leq \chi_m(G) \leq \chi(G) \leq n. \tag{8.1}$$

If u and v are vertices (adjacent or not) of a graph G such that $\deg u \neq \deg v$, then
necessarily $\text{code}_m(u) \neq \text{code}_m(v)$. On the other hand, if G contains two adjacent
vertices u and v with $\deg u = \deg v$, then in order for c to be a multiset coloring, c
must assign at least two distinct colors to the neighbors of u and v. Thus, we have
the following observation from [21].

Observation 8.1.1 ([21]). *The multiset chromatic number of a graph G is 1 if and
only if every two adjacent vertices of G have distinct degrees.*

Since every nonempty bipartite graph has chromatic number 2, the following is
an immediate consequence of (8.1) and Observation 8.1.1.

Proposition 8.1.2 ([21]). *If G is a bipartite graph, then*

$$\chi_m(G) = \begin{cases} 1 \text{ if every two adjacent vertices of } G \text{ have distinct degrees} \\ 2 \text{ otherwise.} \end{cases}$$

The following observation is often useful.

Observation 8.1.3. *If u and v are two adjacent vertices in a graph G such that
$N(u) - \{v\} = N(v) - \{u\}$, then $c(u) \neq c(v)$ for every multiset coloring c of G.*

8.2 Complete Multipartite Graphs

We have noted that for each vertex coloring of a graph G, every two vertices with different degrees have distinct color codes. From this, it follows that determining the multiset chromatic number of G is most interesting and most challenging when G has many vertices of the same degree. We now initiate a study of graphs having this property, especially regular graphs. It is a consequence of Observation 8.1.3 that $\chi_m(K_n) = n$. By (8.1), a graph G of order n has multiset chromatic number n if and only if $G = K_n$.

By Proposition 8.1.2, for the complete bipartite graph $K_{s,t}$,

$$\chi_m(K_{s,t}) = \begin{cases} 1 & \text{if } s \neq t \\ 2 & \text{if } s = t. \end{cases}$$

We now determine the multiset chromatic numbers of all complete multipartite graphs, beginning with the regular complete multipartite graphs, that is, those complete multipartite graphs all of whose partite sets are of the same cardinality. If every partite set of a complete k-partite graph G has n vertices, then we write $G = K_{k(n)}$, where then $K_{k(1)} = K_k$ and $K_{1(n)} = \overline{K}_n$.

For positive integers ℓ and n with $1 \leq n \leq \ell$, let

$$f(\ell, n) = \binom{n + \ell - 1}{\ell - 1}$$

denote the number of n-element submultisets of an ℓ-element set. We now determine the multiset chromatic number of all regular complete multipartite graphs.

Theorem 8.2.1 ([21]). *For positive integers k and n with $1 \leq n \leq \ell$, the multiset chromatic number of the regular complete k-partite graph $K_{k(n)}$ is the unique positive integer ℓ for which*

$$f(\ell - 1, n) < k \leq f(\ell, n).$$

Proof. Denote the partite sets of $G = K_{k(n)}$ by $U_1, U_2, \ldots, U_k$, where then $|U_i| = n$ for each i with $1 \leq i \leq k$. We first claim that $\chi_m(G) \geq \ell$. Assume, to the contrary, that $\chi_m(G) \leq \ell - 1$. Then there exists a multiset $(\ell - 1)$-coloring c of G. Let $A = \{1, 2, \ldots, \ell - 1\}$ denote the set of colors used by c and let S be the set of all n-element multisubsets of the set A. Thus, $|S| = f(\ell - 1, n)$. For $1 \leq i \leq k$, let S_i be the n-element multisubset of A that is used to color the vertices of U_i. Since $k > f(\ell - 1, n)$, it follows that $S_i = S_j$ for some pair i, j of distinct integers with $1 \leq i, j \leq k$. However then, for $u \in U_i$ and $v \in U_j$, it follows that $code(u) = code(v)$, which is impossible. Thus, as claimed, $\chi_m(G) \geq \ell$.

Next, we show that $\chi_m(G) \leq \ell$. Let $B = [\ell] = \{1, 2, \ldots, \ell\}$. Since $k \leq f(\ell, n)$, there exist k distinct submultisets $B_1, B_2, \ldots, B_k$ of B. For each i ($1 \leq i \leq k$), assign the colors in the multiset B_i to the vertices of U_i. Let u and v be two adjacent vertices of G. Then $u \in U_i$ and $v \in U_j$ for distinct integers i and j with $1 \le i, j \le k$. Let B' be the multiset of colors of the vertices in $V(G) - (U_i \cup U_j)$. Since $M(u) = B_j \cup B'$, $M(v) = B_i \cup B'$, and $B_i \ne B_j$, it follows that $M(u) \ne M(v)$. Hence, this ℓ-coloring is a multiset coloring and so $\chi(G) \le \ell$. $\square$

We now consider more general complete multipartite graphs. We denote a complete multipartite graph containing k_i partite sets of cardinality n_i $(1 \le i \le t)$ by $K_{k_1(n_1), k_2(n_2), \dots, k_t(n_t)}$.

Theorem 8.2.2 ([21]). *Let* $G = K_{k_1(n_1), k_2(n_2), \dots, k_t(n_t)}$, *where* $n_1, n_2, \dots, n_t$ *are* t *distinct positive integers. Then*

$$\chi_m(G) = \max\{\chi_m(K_{k_i(n_i)}) : 1 \le i \le t\}.$$

Proof. Let $\ell_i = \chi_m(K_{k_i(n_i)})$ for $1 \le i \le t$. Assume, without loss of generality, that

$$\ell_1 = \max\{\chi_m(K_{k_i(n_i)}) : 1 \le i \le t\}.$$

We first show that $\chi_m(G) \le \ell_1$. For each integer i with $1 \le i \le t$, let c_i be a multiset ℓ_i-coloring of the subgraph $K_{k_i(n_i)}$ in G using the colors in $[\ell_i] = \{1, 2, \dots, \ell_i\}$. We can now define a multiset ℓ_1-coloring c of G by

$$c(x) = c_i(x) \quad \text{if } x \in V(K_{k_i(n_i)}) \text{ for } 1 \le i \le t.$$

Thus, $\chi_m(G) \le \ell_1$. Next, we show that $\chi_m(G) \ge \ell_1$. Assume, to the contrary, that $\chi_m(G) = \ell \le \ell_1 - 1$. Let c' be a multiset ℓ-coloring of G. Then c' induces a coloring c'_1 of the subgraph $K_{k_1(n_1)}$ in G such that $c'_1(x) = c'(x)$ for all $x \in V(K_{k_1(n_1)})$. Since c'_1 uses at most ℓ colors and $\chi_m(K_{k_1(n_1)}) = \ell_1 > \ell$, it follows that c'_1 is not a multiset coloring of $K_{k_1(n_1)}$ and so there exist two adjacent vertices u and v in $K_{k_1(n_1)}$ having the same code with respect to c'_1. Since u and v are both adjacent to every vertex in $V(G) - V(K_{k_1(n_1)})$, it follows that u and v have the same code in G with respect to c', which is a contradiction. $\square$

In particular, if $k_1 = k_2 = \cdots = k_t = 1$, then $K_{k_i(n_i)} = K_{1(n_i)} = \overline{K}_{n_i}$ for $1 \le i \le t$. Since $\chi_m(\overline{K}_{n_i}) = 1$ for $1 \le i \le t$, it follows that $\chi_m(K_{n_1, n_2, \dots, n_t}) = 1$, where $n_1, n_2, \dots, n_t$ are t distinct positive integers.

By (8.1), if G is a graph with $\chi_m(G) = a$ and $\chi(G) = b$, then $a \le b$. In fact, each pair a, b of positive integers with $a \le b$ is realizable as the multiset chromatic number and chromatic number, respectively, for some connected graph.

Proposition 8.2.3 ([21]). *For each pair a, b of positive integers with $a \le b$, there exists a connected graph G such that $\chi_m(G) = a$ and $\chi(G) = b$.*

Proof. If $a = b$, let $G = K_a$ and then $\chi_m(G) = \chi(G) = a$. Thus, we may assume that $a < b$. Let G be a complete b-partite graph with partite sets $V_1, V_2, \dots, V_b$, where $|V_i| = 1$ for $1 \le i \le a$ and $2 \le |V_{a+1}| < |V_{a+2}| < \cdots < |V_b|$.

Then $\chi(G) = b$. It remains to show that $\chi_m(G) = a$. Let $U = V_1 \cup V_2 \cup \cdots \cup V_a$. By Observation 8.1.3, if c is a multiset coloring of G, then $c(x) \neq c(y)$ for every two distinct vertices x and y in U, which implies that $\chi_m(G) \geq a$. On the other hand, the coloring that assigns color i to the vertex in V_i for $1 \leq i \leq a$ and color 1 to the remaining vertices of G is a multiset a-coloring of G. Therefore, $\chi_m(G) = a$. □

8.3 Graphs with Prescribed Order and Multiset Chromatic Number

If G is a connected graph of order n and $\chi_m(G) = k$, then $1 \leq k \leq n$. Furthermore, $\chi_m(G) = n$ if and only if $G = K_n$. For nearly every pair k, n of positive integers with $k \leq n$, there is a connected graph G of order n having multiset chromatic number k.

Proposition 8.3.1 ([21]). *Let k and n be integers with $1 \leq k \leq n$. Then there exists a connected graph G of order n with $\chi_m(G) = k$ if and only if $k \neq n - 1$.*

Proof. For $n = 1, 2$, the result immediately follows. Hence, suppose that $n \geq 3$. For $k = 1$, let G be a connected graph of order n such that no two adjacent vertices of G have the same degree. Then $\chi_m(G) = 1$. For $k = n$, let $G = K_n$ and so $\chi_m(G) = n$. For $2 \leq k \leq n - 2$, let $G = K_{1,1,\ldots,1,n-k}$ be the complete $(k + 1)$-partite graph such that k partite sets of G are singletons and one partite set of G consists of $n - k$ vertices. Since $n - k \geq 2$, it follows that $\chi_m(G) = k$. For the converse, assume, to the contrary, that there is a connected graph G of order n with $\chi_m(G) = n - 1$. Then $G \neq K_n$ and $\chi(G) = n - 1$. Thus, G is obtained from K_{n-1} by joining a new vertex to some (but not all) vertices of K_{n-1}. Let $V(G) = \{v_1, v_2, \ldots, v_n\}$, where the subgraph induced by $V(G) - \{v_n\}$ is K_{n-1} and v_n is adjacent to $v_1, v_2, \ldots, v_t$, where $1 \leq t \leq n - 2$. The $(n - 2)$-coloring c of G given by

$$c(v_i) = \begin{cases} i & \text{if } 1 \leq i \leq t \\ i - 1 & \text{if } t + 1 \leq i \leq n - 1 \\ n - 2 & \text{if } i = n \end{cases}$$

is a multiset coloring and so $\chi_m(G) \leq n - 2$, which is a contradiction. □

By Proposition 8.3.1, $\chi_m(G) \leq n - 2$ if and only if $G \neq K_n$. In fact, there are only two connected graphs of order $n \geq 6$ having multiset chromatic number $n - 2$. In order to show this, we first prove a useful lemma. For two graphs F and H, recall that the *union* of F and H is denoted by $F + H$, while the *join* of F and H is denoted by $F \vee H$.

Lemma 8.3.2. *If G is a connected graph of order $n \geq 6$ and $\Delta(G) \leq n - 2$, then*

$$\chi_m(G) \leq n - 3.$$

Proof. Since G is connected and $\Delta(G) \leq n-2$, the complement $\overline{G}$ of G contains $2K_2$ as a subgraph. If $\overline{G}$ contains either $K_2 + K_3$ or $3K_2$ as a subgraph, then $\chi(G) \leq n-3$ and so $\chi_m(G) \leq n-3$. Otherwise, let u_1, u_2, w_1 and w_2 be four distinct vertices in G such that $u_1w_1, u_2w_2 \notin E(G)$ and

$$X = V(G) - \{u_1, u_2, w_1, w_2\} = \{v_1, v_2, \ldots, v_{n-4}\}.$$

Since $\overline{G}$ does not contain $3K_2$, it follows that the subgraph induced by the $n-4$ vertices in X is K_{n-4}.

Suppose that there exists a vertex $v \in X$ that is adjacent to both u_1 and w_1 or to both u_2 and w_2, say v_1 is adjacent to both u_1 and w_1. Then the $(n-3)$-coloring $c_1 : V(G) \to \{1, 2, \ldots, n-3\}$ given by

$$c_1(x) = \begin{cases} i & \text{if } x = v_i \ (1 \leq i \leq n-4) \\ 1 & \text{if } x = u_1, w_1 \\ n-3 & \text{if } x = u_2, w_2 \end{cases}$$

is proper. Therefore, $\chi_m(G) \leq n-3$.

Only one case remains to be considered. For each $i = 1, 2$, suppose that one of u_i and w_i is adjacent to every vertex in X and the other is adjacent to no vertex in X, say u_1 and u_2 are adjacent to every vertex in X and w_1 and w_2 are adjacent to no vertex in X. Therefore, $\deg v = n-3$ for every $v \in X$, while

$$\deg u_i \in \{n-4, n-3, n-2\} \quad \text{and} \quad \deg w_i \in \{1, 2\}.$$

Also, $\deg u_i > \deg w_j$ for $1 \leq i, j \leq 2$ and $|\deg u_1 - \deg u_2| \leq 1$. If $\deg u_1 = \deg u_2$, then $u_1w_2, u_2w_1 \in E(G)$. Consider the coloring $c_2 : V(G) \to \{1, 2, \ldots, n-3\}$ defined by

$$c_2(x) = \begin{cases} i & \text{if } x = v_i \ (1 \leq i \leq n-4) \text{ or } x = w_i \ (i = 1, 2) \\ n-3 & \text{if } x = u_1, u_2. \end{cases}$$

If $\deg u_1 \neq \deg u_2$, then let $u \in \{u_1, u_2\}$ such that $\deg u = n-3$ and consider the coloring $c_3 : V(G) \to \{1, 2, \ldots, n-3\}$ defined by

$$c_3(x) = \begin{cases} i & \text{if } x = v_i \ (1 \leq i \leq n-4) \\ n-3 & \text{if } x = u \\ 1 & \text{otherwise.} \end{cases}$$

Then both c_2 and c_3 are multiset colorings and $\chi_m(G) \leq n-3$ in each case.   □

Theorem 8.3.3 ([21]). *If G is a connected graph of order $n \geq 6$, then*

$$\chi_m(G) = n-2 \text{ if and only if } G \text{ is either } K_n - e \text{ or } (K_{n-2} + K_1) \vee K_1.$$

Proof. Let G be a connected graph of order $n \geq 6$. It is clear that if G is either $K_n - e$ or $(K_{n-2} + K_1) \vee K_1$, then $\chi_m(G) = n - 2$.

For the converse, suppose that $\chi_m(G) = n - 2$ and let c be a multiset $(n-2)$-coloring of G. Then $G \neq K_n$ and by Lemma 8.3.2, $\Delta(G) = n - 1$. Let $X = \{v_1, v_2, \ldots, v_{n'}\}$ be the set of vertices in G of degree $n - 1$ and $Y = V(G) - X$. (Hence, $1 \leq n' \leq n - 2$.) Thus, c must assign a unique color to each vertex in X. Let H be the subgraph induced by the $n - n'$ vertices in Y. Hence,

$$n - 2 = \chi_m(G) \leq \max\{n', \chi_m(H)\}.$$

Since $G \neq K_n$, it follows that $H \neq K_{n-n'}$.

If $n' = n - 2$, then $H = 2K_1$ and $G = K_n - e$. If $n' \leq n - 3$, then let $H_1, H_2, \ldots, H_s$ be the components of H, where each H_i is a graph of order n_i and $n_1 \geq n_2 \geq \cdots \geq n_s$. Hence,

$$n - 2 \leq \chi_m(H) = \max\{\chi_m(H_i) : 1 \leq i \leq s\} \leq n_1 \leq n - s,$$

that is, $s = 1$ or $s = 2$. If $s = 1$, then H is a noncomplete connected graph of order $n - n'$ and so $\chi_m(H) \leq (n - n') - 2 < n - 2$, which is impossible. If $s = 2$, then $\chi_m(H) = n_1 = n - 2$. Hence, $H_1 = K_{n-2}$ and $H_2 = K_1$, implying that $G = (K_{n-2} + K_1) \vee K_1$. □

8.4 Multiset Colorings Versus Set Colorings

Since a set coloring of a connected graph G is a multiset coloring of G, it follows that

$$\chi_m(G) \leq \chi_s(G).$$

Next, we show that every pair a, b of positive integers with $a \leq b$ can be realized as the multiset chromatic number and set chromatic number, respectively, of some connected graph. Recall the lower bound for the set chromatic number $\chi_s(G)$ of a graph G (Theorem 7.3.2), namely,

$$\chi_s(G) \geq 1 + \lceil \log_2 \omega(G) \rceil. \tag{8.2}$$

Theorem 8.4.1 ([21]). *For each pair a, b of positive integers with $a \leq b$, there exists a connected graph G such that $\chi_m(G) = a$ and $\chi_s(G) = b$.*

Proof. If $a = b$, then the complete graph K_a has the desired property. Thus, we may assume that $a < b$. We consider two cases, according to whether $a = 1$ or $a \geq 2$.

Case 1. $a = 1$. Then $b \geq 2$. We show that there is a graph H such that $\chi_m(H) = 1$ and $\chi_s(H) = b \geq 2$. We begin by constructing a graph F. Let $S_1, S_2, \ldots, S_{2^{b-1}}$ be the 2^{b-1} subsets of $[b-1]$, where $|S_1| \leq |S_2| \leq \cdots \leq |S_{2^{b-1}}|$. Hence, $S_1 = \emptyset$ and $S_{2^{b-1}} = [b-1]$. Then the graph F is obtained from $K_{2^{b-1}}$ with

$$V(K_{2^{b-1}}) = U = \{u_1, u_2, \ldots, u_{2^{b-1}}\}$$

by adding pairwise disjoint sets $W_2, W_3, \ldots, W_{2^{b-1}}$ to $K_{2^{b-1}}$, where

$$W_i = \{w_{i,1}, w_{i,2}, \ldots, w_{i,|S_i|}\},$$

and joining each vertex in W_i to u_i for each i with $2 \leq i \leq 2^{b-1}$. Since $|S_i| \leq i-1$ for $1 \leq i \leq 2^{b-1}$, we can add more pendant edges at each vertex u_i, if necessary, to obtain the graph H such that $\deg_H u_i = 2^{b-1} - 2 + i$ for $1 \leq i \leq 2^{b-1}$. Figure 8.2 shows the graph H for $b = 4$. Let $X = V(H) - V(F)$. For $b = 4$, the set X consists of the solid vertices shown in Fig. 8.2, while the vertex set $V(F)$ of F consists of all open vertices.

Since every two adjacent vertices in H have different degrees, $\chi_m(H) = 1$ by Observation 8.1.1. It remains only to show that $\chi_s(H) = b$. Because $\omega(H) = 2^{b-1}$, it follows by (8.2) that $\chi_s(H) \geq b$. On the other hand, consider the coloring $c_1 : V(H) \to [b]$ of H that assigns (i) the color b to each vertex in $U \cup X$ and (ii) the colors in S_i to the $|S_i|$ end-vertices in W_i for $2 \leq i \leq 2^{b-1}$. Figure 8.2 shows such a coloring for $b = 4$. Then $c(N(u_i)) = S_i \cup \{b\}$ for $1 \leq i \leq 2^{b-1}$. Since $|c(N(u_i))| \geq 2$ for $2 \leq i \leq 2^{b-1}$ and $|c(N(x))| = 1$ for each end-vertex in H, it follows that c_1 is a set b-coloring. Therefore, $\chi_s(H) = b$.

Case 2. $a \geq 2$. Then $b \geq 3$. We now construct a graph G from the graph H in Case 1 and the complete graph K_a with $V(K_a) = Y = \{y_1, y_2, \ldots, y_a\}$ by joining each vertex y_i to the vertex $w_{2,1}$ in H for $1 \leq i \leq a$ (see Fig. 8.3 for $a = 3$ and $b = 4$). Now, two vertices are adjacent and have the same degree if and only if both vertices belong to Y. Therefore, no multiset coloring can assign the same color to two distinct vertices in Y and so $\chi_m(G) \geq |Y| = a$. Next, consider the coloring that assigns (i) the color i to the vertex y_i in Y for $1 \leq i \leq a$ and (ii) the color 1 to the remaining vertices. Since this coloring is a multiset a-coloring

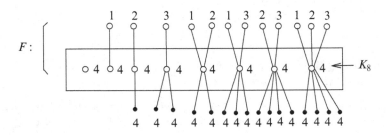

Fig. 8.2 The graph H in Case 1 for $a = 1$ and $b = 4$

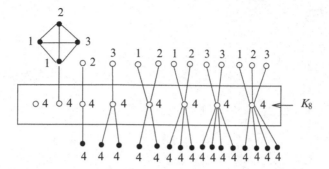

Fig. 8.3 The graph G in Case 2 of the proof of Theorem 8.4.1 for $a = 3$ and $b = 4$

of G, it follows that $\chi_m(G) = a$. To verify that $\chi_s(G) = b$, observe first that $\chi_s(G) \geq b$ again by (8.2). On the other hand, the coloring $c_2 : V(G) \rightarrow [b]$ such that c_2 restricted to $V(H)$ is the coloring c_1 mentioned above and $c_2(y_i) = i$ for $1 \leq i \leq a$ is a set b-coloring of G. Figure 8.3 shows such a coloring for $a = 3$ and $b = 4$. Thus, $\chi_s(G) = b$, as desired.   □

Chapter 9
Metric Colorings

In this chapter we describe yet another proper vertex coloring induced by a given nonproper vertex coloring of a graph. This proper vertex coloring is defined with the aid of distances and this too may very well require fewer colors than the chromatic number of the graph.

9.1 Metric Chromatic Number

Recall that the *distance* $d(u, v)$ between two vertices u and v in a connected graph G is the length of a shortest $u - v$ path and the *diameter* diam(G) of G is the greatest distance between two vertices of G. For a set $S \subseteq V(G)$ and a vertex v of G, the *distance $d(v, S)$ between v and S* is defined as

$$d(v, S) = \min\{d(v, x) : x \in S\}.$$

Then $0 \le d(v, S) \le \text{diam}(G)$, where $d(v, S) = 0$ if and only if $v \in S$. Suppose that $c : V(G) \to [k]$ is a k-coloring of G for some positive integer k where adjacent vertices may be colored the same and let $V_1, V_2, \ldots, V_k$ be the resulting color classes. A k-vector, called the *metric color code*, can be associated with each vertex v of G, which is denoted by $\text{code}_\mu(v)$ and defined by

$$\text{code}_\mu(v) = (a_1, a_2, \cdots, a_k) = a_1 a_2 \cdots a_k,$$

where for each integer i with $1 \le i \le k$, $a_i = d(v, V_i)$. (We write $\text{code}_{\mu,c}(v)$ to indicate the metric color code of a vertex v with respect to a specific coloring c.) If $\text{code}_\mu(u) \ne \text{code}_\mu(v)$ for every two adjacent vertices u and v of G, then c is called a *metric coloring* of G. The minimum k for which G has a metric k-coloring is called the *metric chromatic number* of G and is denoted by $\mu(G)$. These concepts were

© The Author 2016
P. Zhang, *A Kaleidoscopic View of Graph Colorings*, SpringerBriefs in Mathematics,
DOI 10.1007/978-3-319-30518-9_9

Fig. 9.1 A 4-chromatic
graph G with $\mu(G) = 3$

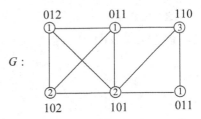

$$G:$$

introduced and studied in [22]. For example, the graph G of Fig. 9.1 has chromatic number 4 and metric chromatic number 3. A metric 3-coloring of G is shown in Fig. 9.1 together with the metric color code of each vertex of G.

The following observations will be useful to us.

Observation 9.1.1. *To show that a given coloring c is a metric coloring, it suffices to show that* $\text{code}_\mu(u) \neq \text{code}_\mu(v)$ *for adjacent vertices u and v with $c(u) = c(v)$.*

Observation 9.1.2. *Let c be a metric coloring of a connected graph G and $uv \in E(G)$. If $d(u,w) = d(v,w)$ for all $w \in V(G) - \{u,v\}$, then $c(u) \neq c(v)$.*

9.2 Graphs with Prescribed Order and Metric Chromatic Number

Let c be a proper k-coloring of a nontrivial connected graph G with resulting color classes $V_1, V_2, \ldots, V_k$ and let u and v be two adjacent vertices of G. Then $u \in V_i$ and $v \in V_j$ for some $i, j \in \{1, 2, \ldots, k\}$ with $i \neq j$. Suppose that $\text{code}_\mu(u) = (a_1, a_2, \cdots, a_k)$ and $\text{code}_\mu(v) = (b_1, b_2, \cdots, b_k)$. Then $a_i = b_j = 0$ and $a_j = b_i = 1$. Thus, $\text{code}_\mu(u) \neq \text{code}_\mu(v)$ and so c is also a metric coloring of G. Consequently,

$$2 \leq \mu(G) \leq \chi(G) \leq n \tag{9.1}$$

for every nontrivial connected graph G of order n. It is obvious that a connected graph G of order n has metric chromatic number n if and only if $G = K_n$. As is the case with proper colorings, only bipartite graphs have metric 2-colorings.

Theorem 9.2.1 ([22]). *A nontrivial connected graph G has metric chromatic number 2 if and only if G is bipartite.*

Proof. If G is bipartite, then $\chi(G) = 2$ and so the result follows by (9.1). For the converse, suppose that there is a connected graph G with metric chromatic number 2 that is not bipartite. Let $C = (v_1, v_2, \ldots, v_\ell, v_{\ell+1} = v_1)$ be an odd cycle in G and let c be a metric 2-coloring of G. Consider the (cyclic) color sequence $s : c(v_1), c(v_2), \ldots, c(v_\ell), c(v_{\ell+1}) = c(v_1)$. By a *block* of s, we mean a maximal subsequence of s consisting of terms of the same color. We claim that each block of s must be of odd length. Suppose, to the contrary, that there is a block of even length, say $c(v_1), c(v_2), \ldots, c(v_{2k})$ for some positive integer k where $2k < \ell$. We may assume that $c(v_i) = 1$ for $1 \leq i \leq 2k$ and so $c(v_\ell) = c(v_{2k+1}) = 2$. Thus, $\mathrm{code}_\mu(v_i) = (0, d_i)$, where d_i is the distance between v_i and the nearest vertex colored 2. Since $\mathrm{code}_\mu(v_1) = (0, 1)$ and $\mathrm{code}_\mu(v_i) \neq \mathrm{code}_\mu(v_{i+1})$ for $1 \leq i \leq 2k - 1$, it follows that $|d_i - d_{i+1}| = 1$. Consequently, d_i is odd if and only if i is odd for $1 \leq i \leq 2k - 1$. However, $\mathrm{code}_\mu(v_{2k}) = (0, 1)$, producing a contradiction. Thus, as claimed, every block of s has odd length. Hence, we may assume that either (i) s consists of a single block in which $c(v_i) = 1$ for $1 \leq i \leq \ell$ or (ii) s contains an even number of blocks, each having an odd number of terms. If (i) occurs, then $\mathrm{code}_\mu(v_i) = (0, d_i)$ for some $d_i \geq 1$ and $|d_i - d_{i+1}| = 1$ for $1 \leq i \leq \ell$, which implies that $d_1, d_2, \ldots, d_\ell, d_1$ alternate between even and odd integers, which is impossible since ℓ is odd. If (ii) occurs, then ℓ is even, which contradicts our assumption.   □

The following is a consequence of Theorem 9.2.1.

Corollary 9.2.2 ([22]). *If G is a connected graph with $\chi(G) = 3$, then $\mu(G) = 3$.*

By Proposition 9.2.1, if G is a complete bipartite graph, then $\mu(G) = \chi(G) = 2$; while by Corollary 9.2.2, if G is a complete 3-partite graph, then $\mu(G) = \chi(G) = 3$. This fact can be extended to all complete multipartite graphs, as we show next.

Proposition 9.2.3 ([22]). *If G is a complete k-partite graph where $k \geq 2$, then*

$$\mu(G) = k.$$

Proof. We proceed by induction on $k \geq 2$. For $k = 2$, the result follows by Proposition 9.2.1. Suppose that the metric chromatic number of every complete $(k - 1)$-partite graph is $k - 1$ for some integer $k \geq 3$. Let G be a complete k-partite graph with partite sets $U_1, U_2, \ldots, U_k$. Certainly, the metric chromatic number of G is at most k. Assume, to the contrary, that $\mu(G) \leq k - 1$. Let there be given a metric $(k - 1)$-coloring c of G using the colors in $[k - 1]$. We claim that for each partite set U_i ($1 \leq i \leq k$), the coloring c_i that is the restriction of c to $V(G) - U_i$ in the complete $(k - 1)$-partite graph $G - U_i$ is a metric coloring. To see this, let u and v be two adjacent vertices in $G - U_i$ where $1 \leq i \leq k$. Then $\mathrm{code}_{\mu,c}(u) \neq \mathrm{code}_{\mu,c}(v)$ Since $d(u, x) = d(v, x) = 1$ for all $x \in U_i$, it follows that $\mathrm{code}_{\mu,c_i}(u) \neq \mathrm{code}_{\mu,c_i}(v)$ This implies that c_i is a metric coloring of $G - U_i$. Since $\mu(G - U_i) = k - 1$ by the induction hypothesis, it follows that $c(V(G) - U_i) = [k - 1]$ for every partite set U_i of G. Because $c(V(G)) = [k - 1]$, there are vertices x and y in G belonging to different partite sets of G such that $c(x) = c(y)$, say $c(x) = c(y) = 1$. We may assume that $x \in U_1$ and $y \in U_2$. Since $c(V(G) - U_1) = c(V(G) - U_2) = [k - 1]$, it follows that $\mathrm{code}_{\mu,c}(x) = \mathrm{code}_{\mu,c}(y) = (0, 1, 1, \cdots, 1)$, a contradiction.   □

Since the metric chromatic number of the complete k-partite graph $K_{1,1,\ldots,1,n-(k-1)}$ of order n is k, the following result is a consequence of Proposition 9.2.3.

Corollary 9.2.4 ([22]). *For each pair k, n of integers with $2 \leq k \leq n$, there is a connected graph G of order n with $\mu(G) = k$.*

We have seen that a connected graph G of order n has metric chromatic number n if and only if $G = K_n$. We next determine those connected graphs of order $n \geq 3$ having metric chromatic number $n - 1$.

Theorem 9.2.5 ([22]). *A connected graph G of order $n \geq 3$ has metric chromatic number $n - 1$ if and only if*

$$G = K_{n-2} \vee \overline{K}_2 \text{ or } G = (K_{n-2} + K_1) \vee K_1.$$

Proof. It is easy to see that $\mu(G) = n - 1$ if $G = K_{n-2} \vee \overline{K}_2$ or $G = (K_{n-2} + K_1) \vee K_1$. For the converse, assume that G is a connected graph of order $n \geq 3$ with $\mu(G) = n - 1$. Thus, $G \neq K_n$ and $\chi(G) = n - 1$. Then the clique number of G is $\omega(G) = n - 1$. Let $H = K_{n-1}$ be a clique of order $n - 1$ in G with $V(H) = \{v_1, v_2, \ldots, v_{n-1}\}$ and let $V(G) - V(H) = \{v\}$. If v is adjacent to exactly $n - 2$ vertices of H, then $G = K_{n-2} \vee \overline{K}_2$; while if v is adjacent to exactly one vertex of H, then $G = (K_{n-2} + K_1) \vee K_1$. Thus, we may assume that v is adjacent to $v_1, v_2, \ldots, v_k$ in G where $2 \leq k \leq n - 3$. Define a coloring c of G by $c(v_i) = i$ for $1 \leq i \leq n - 3$, $c(v_{n-2}) = 1$, $c(v_{n-1}) = 2$ and $c(v) = n - 2$. Since v_1 and v_2 are adjacent to v and v_{n-2} and v_{n-1} are not, it follows that $\text{code}_\mu(v_1) \neq \text{code}_\mu(v_{n-2})$ and $\text{code}_\mu(v_2) \neq \text{code}_\mu(v_{n-1})$. By Observation 9.1.1, c is a metric $(n-2)$-coloring of G and so $\mu(G) \leq n - 2$, which contradicts our assumption. $\square$

9.3 Bounds for the Metric Chromatic Number of a Graph

We have already noted that if G is a nontrivial connected graph of order n, then

$$2 \leq \mu(G) \leq \chi(G) \leq n.$$

With the aid of other graphical parameters, we now present some improved lower and upper bounds for $\mu(G)$. It is well known that $\chi(G) \geq \omega(G)$ for every graph G. It need not occur that $\mu(G) \geq \omega(G)$ however. For the graph G of Fig. 9.2, $\chi(G) = \omega(G) = 4$. Since the 3-coloring of G shown in Fig. 9.2 is a metric coloring, it follows

Fig. 9.2 A graph G with $\mu(G) = 3$ and $\omega(G) = 4$

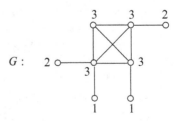

that $\mu(G) \leq 3$ and so $\mu(G) = 3$ by Proposition 9.2.1. There is a lower bound for the metric chromatic number of a graph in terms of its clique number, however. This bound is presented next.

Theorem 9.3.1 ([22]). *For every nontrivial connected graph G,*

$$\mu(G) \geq 1 + \lceil \log_2 \omega(G) \rceil.$$

Proof. Let H be a clique of order $\omega = \omega(G)$ in G with $V(H) = \{v_1, v_2, \ldots, v_\omega\}$. Suppose that $\mu(G) = k$ and c is a metric k-coloring of G using colors in the set $[k]$. Suppose that $|c(V(H))| = r$, say $c(V(H)) = [r]$, where $1 \leq r \leq k$.

Let $v \in V(H)$ with $\text{code}_\mu(v) = (a_1, a_2, \cdots, a_k)$. Observe that $a_i \in \{0, 1\}$ for $1 \leq i \leq r$ and exactly one of $a_1, a_2, \ldots, a_r$ is 0. Thus, there are r possibilities for the r-tuple $(a_1, a_2, \ldots, a_r)$. For each j with $r + 1 \leq j \leq k$, let d_j be the minimum distance between a vertex colored j in $V(G) - V(H)$ and a vertex in $V(H)$. Then $a_j \in \{d_j, d_j + 1\}$ and so there are two choices for each coordinate a_j when $r + 1 \leq j \leq k$. Thus, there are 2^{k-r} possibilities for the $(k-r)$-tuple $(a_{r+1}, a_{r+2}, \ldots, a_k)$. Therefore, there are $r \cdot 2^{k-r}$ possible metric color codes for the vertices of H and so $\omega \leq r \cdot 2^{k-r}$. Since $r \leq 2^{r-1}$ for each positive integer r, it follows that $\omega \leq 2^{r-1} \cdot 2^{k-r} = 2^{k-1}$ and so $k - 1 \geq \log_2 \omega$. Therefore, $k = \mu(G) \geq 1 + \log_2 \omega$, producing the desired result. $\square$

The lower bound for the metric chromatic number of a graph in Theorem 9.3.1 is sharp. Consider the graph G of order 11 shown in Fig. 9.3 consisting of a complete subgraph H of order 8, where

$$V(H) = \{v_{ijk} : i, j, k \in \{0, 1\}\},$$

and three additional vertices x, y and z, where x is adjacent to v_{ijk} if and only if $i = 1$, y is adjacent to v_{ijk} if and only if $j = 1$ and z is adjacent to v_{ijk} if and only if $k = 1$. (Many edges of G belonging to the subgraph H have been omitted in Fig. 9.3.) By Theorem 9.3.1, $\mu(G) \geq 1 + \lceil \log_2 \omega(G) \rceil = 4$. Since the 4-coloring defined by $c(x) = 1$, $c(y) = 2$, $c(z) = 3$ and $c(v_{ijk}) = 4$ for all $i, j, k \in \{0, 1\}$ is a metric coloring, it follows that $\mu(G) = 4$.

Fig. 9.3 An 8-chromatic graph G with $\mu(G) = 1 + \lceil \log_2 \omega(G) \rceil = 4$

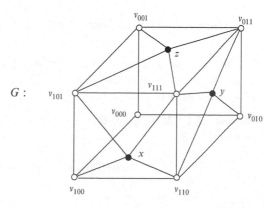

If G is a connected graph with $\text{diam}(G) = d$ and $\mu(G) = k$, then for a metric k-coloring of G and a vertex v of G, exactly one coordinate of $\text{code}_\mu(v)$ is 0, while each of the remaining coordinates of $\text{code}_\mu(v)$ is an element of the set $\{1, 2, \ldots, d\}$, resulting in at most $k \cdot d^{k-1}$ distinct metric color codes for the vertices of G. Note that each metric k-coloring c of a graph G provides a proper coloring c^* of G, where $c^*(v) = \text{code}_\mu(v)$ for each $v \in V(G)$. Since c^* uses at most $k \cdot d^{k-1}$ different colors, $\chi(G) \leq k \cdot d^{k-1}$. This observation is stated below.

Observation 9.3.2 ([22]). *If G is a connected graph of order n, diameter d and metric chromatic number k, then*

$$\chi(G) \leq k \cdot d^{k-1}.$$

Since $\chi(G) \leq n - d + 1$ for every connected graph G of order n and diameter d by Theorem 1.2.7, we obtain the following result that presents an upper bound for the metric chromatic number of a graph in terms of its order and diameter.

Proposition 9.3.3 ([22]). *If G is a nontrivial connected graph of order n and diameter d, then*

$$\mu(G) \leq n - d + 1.$$

The upper bound in Proposition 9.3.3 is sharp. To see this, we construct a connected graph G of order n and diameter d such that $\mu(G) = n - d + 1$ for each pair n, d of integers with $1 \leq d \leq n - 1$. Let G be the graph obtained from the graph K_{n-d+1} and the path $P_d = (v_1, v_2, \ldots, v_d)$ by identifying a vertex of K_{n-d+1} and the vertex v_1 of P_d and denoting the identified vertex by v_1. Then the order of G is n and the diameter of G is d. It remains to show that $\mu(G) = n - d + 1$. By Proposition 9.3.3, $\mu(G) \leq n - d + 1$. By Observation 9.1.2, the $n - d$ vertices in $V(K_{n-d+1}) - \{v_1\}$ must be assigned different colors in every metric coloring of G and so $\mu(G) \geq n - d$. Assume, to the contrary, that there is a metric $(n - d)$-coloring c of G. Then $c(v_1) = c(x)$ for some $x \in V(K_{n-d+1}) - \{v_1\}$. However then, $\text{code}_\mu(v_1) = \text{code}_\mu(x)$, which is impossible. Thus, $\mu(G) = n - d + 1$, as claimed.

9.4   Metric Colorings Versus Other Colorings

While the removal of a vertex from a given graph can never result in a graph with a larger chromatic number than that of the given graph, this is not the case for the metric chromatic number.

Theorem 9.4.1 ([22]). *If v is a vertex that is not a cut-vertex of a connected graph G, then*

$$\mu(G - v) \leq \mu(G) + \deg v.$$

Fig. 9.4 A graph G and a vertex v of G with $\mu(G - v) = \mu(G) + \deg v$

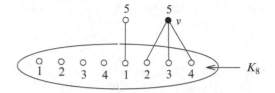

The upper bound for $\mu(G - v)$ in Theorem 9.4.1 is sharp. For example, Fig. 9.4 shows a graph G and a vertex v with $\deg v = 3$ such that $\mu(G) = 5$ and $\mu(G-v) = 8 = \mu(G) + \deg v$. A metric 5-coloring of G is shown in Fig. 9.4 as well.

Similar to set colorings, each pair a, b of integers with $2 \leq a \leq b$ can be realized as the metric chromatic number and chromatic number, respectively, of a connected graph under some restrictions for b in terms of a.

Theorem 9.4.2 ([22]). *For each pair a, b of integers with $2 \leq a \leq b \leq 2^{a-1}$, there exists a connected graph G with $\mu(G) = a$ and $\chi(G) = b$.*

Proof. If $a = b$, then $\mu(K_b) = \chi(K_b) = b$. Hence, we may assume that $3 \leq a < b \leq 2^{a-1}$. Consider the function

$$f_a : [0, a - 2] \to [a, 2^{a-1}]$$

defined by

$$f_a(x) = 2^x(a - x).$$

Note that f_a is strictly increasing on $[0, a - 2]$. Consequently, there exists an integer $p \in [1, a - 2]$ such that

$$a = f_a(0) \leq f_a(p - 1) < b \leq f_a(p) \leq f_a(a - 2) = 2^{a-1}.$$

Also, observe that

$$p \leq 2^{p-1} < \frac{f_a(p - 1)}{a - p} < \left\lceil \frac{b}{a - p} \right\rceil \leq 2^p.$$

Let $q = \left\lceil \frac{b}{a-p} \right\rceil$. Then $p + 1 \leq q \leq 2^p$ and

$$b = (q - 1)(a - p) + r, \tag{9.2}$$

where $1 \leq r \leq a - p$.

Let $H = K_b$ where $V(H)$ can be partitioned into q subsets $X_1, X_2, \ldots, X_q$ such that $|X_i| = a - p$ for $1 \leq i \leq q - 1$ and $|X_q| = r$. Write

$$X_i = \{x_{i,j} : 1 \leq j \leq |X_i|\}$$

for $1 \leq i \leq q$. Also, let $S_1, S_2, \ldots, S_{2^p}$ be the 2^p subsets of $[p]$, where

$$|S_1| \leq |S_2| \leq \cdots \leq |S_{2^p}|.$$

(Hence, $S_1 = \emptyset$, $|S_i| = 1$ for $2 \leq i \leq p+1$ and $S_{2^p} = [p]$.) Let Y be a set of $\sum_{i=1}^{p} i \binom{p}{i}$ vertices disjoint from $V(H)$ such that Y can be partitioned into $2^p - 1$ subsets $Y_2, Y_3, \ldots, Y_{2^p}$ for which $|Y_i| = |S_i|$ for $2 \leq i \leq 2^p$.

A graph G is constructed from H by

 (i) adding the vertices in $\cup_{i=2}^{q} Y_i$ to H and
 (ii) joining each vertex in Y_i to every vertex in X_i for $2 \leq i \leq q$.

Figure 9.5 shows the graphs G in the case when $(a, b) = (7, 30)$ and $(a, b) = (10, 30)$.

Since $\chi(G) = b$, it remains to show $\mu(G) = a$. The coloring $c_1 : V(G) \to [a]$ defined by

$$c_1(x_{i,j}) = p + j \quad \text{for } 1 \leq i \leq q-1 \text{ and } 1 \leq j \leq a-p$$

$$c_1(x_{q,j}) = p + j \quad \text{for } 1 \leq j \leq r$$

$$c_1(Y_i) = S_i \quad\quad \text{for } 2 \leq i \leq q$$

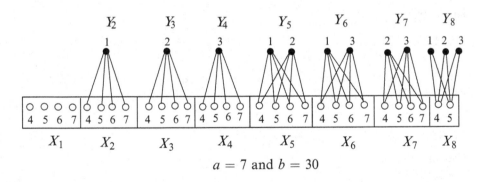

$a = 7$ and $b = 30$

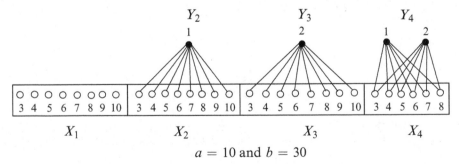

$a = 10$ and $b = 30$

Fig. 9.5 Graphs in the proof of Theorem 9.4.2 for $a \in \{7, 10\}$ and $b = 30$

is a metric a-coloring of G and so $\mu(G) \le a$. Assume, to the contrary, that $\mu(G)$ $< a$. Then there exists a metric $(a-1)$-coloring of G. Permuting the colors $1, 2, \ldots,$ $a - 1$, if necessary, we obtain an $(a - 1)$-coloring $c : V(G) \to [a - 1]$ of G such that $c(X) = [\ell]$, where then $\ell \le a - 1$ and $X = V(H)$. By Observation 9.1.2, the $a - p$ vertices in X_1 must be colored differently by c. Therefore, $a - p \le \ell \le a - 1$. Observe that the first ℓ coordinates of the code of each vertex in X are all 1 except one coordinate is 0. Furthermore, each of the remaining $a - 1 - \ell$ coordinates is either 1 or 2. Therefore,

$$b \le 2^{a-1-\ell} \cdot \ell = 2^{a-1} \left(2^{-\ell} \cdot \ell \right).$$

The function $g(x) = 2^{-x} \cdot x$ defined on $\mathbb{R}$ is decreasing on $\left(\frac{1}{\ln 2}, \infty \right)$ and $\frac{1}{\ln 2} < 2 \le$ $a - p$, implying that

$$b \le 2^{a-1} \left(2^{-\ell} \cdot \ell \right) \le 2^{a-1} \left[2^{-(a-p)} \cdot (a-p) \right] = 2^{p-1}(a-p).$$

However then,

$$2^{p-1}(a - p + 1) = f_a(p - 1) < b \le 2^{p-1}(a - p),$$

which is a contradiction. Therefore, we conclude that $\mu(G) = a$. □

It is not known whether there exists a graph G with $\mu(G) = a$ and $\chi(G) = b$ when $a \ge 3$ and $b > 2^{a-1}$. However, if such a graph G does exist, then it follows by Proposition 9.3.1 that $\omega(G) < b$. In particular, it is not known if there is a 5-chromatic graph whose metric chromatic number is 3.

If c is a set k-coloring of a connected graph G, then $c(N(x)) \ne c(N(y))$ for every two adjacent vertices x and y of G. Thus, there is some color i that belongs to exactly one of $c(N(x))$ and $c(N(y))$. This implies that $\text{code}_\mu(x)$ and $\text{code}_\mu(y)$ differ in the i-th coordinate and so $\text{code}_\mu(x) \ne \text{code}_\mu(y)$. Thus, c is also a metric coloring and so

$$\mu(G) \le \chi_s(G). \tag{9.3}$$

Therefore, if G is a bipartite graph or a complete graph, then $\mu(G) = \chi_s(G)$. It was shown in [22] that there is an infinite class of graphs G for which $\mu(G) < \chi_s(G)$. In fact, more can be said.

Theorem 9.4.3 ([22]). *For each nonnegative integer ℓ, there exists a connected graph G such that $\chi_s(G) - \mu(G) = \ell$.*

We have seen that if G is a graph with $\mu(G) = a$, $\chi_s(G) = b$ and $\chi(G) = c$, then $a \le b \le c$. It was shown in [25] that there is an infinite class of graphs G for which

$$\mu(G) < \chi_s(G) < \chi(G).$$

Theorem 9.4.4 ([25]). *For each integer $k \geq 3$, there exists a connected graph G such that $\mu(G) = k$, $\chi_s(G) = 2^{k-2} + k - 1$ and $\chi(G) = 2^{k-1}$.*

For every connected graph G, we know that

$$\chi_m(G) \leq \chi_s(G) \leq \chi(G) \text{ and } \mu(G) \leq \chi_s(G) \leq \chi(G).$$

However, there is the lingering question of the relationship between $\mu(G)$ and $\chi_m(G)$. Observe that not every metric coloring is a multiset coloring. For example, for the path $P_5 = (v_1, v_2, v_3, v_4, v_5)$ of order 5, the 2-coloring c with $c(v_1) = c(v_2) = c(v_3) = c(v_4) = 1$ and $c(v_5) = 2$ is a metric coloring which is not a multiset coloring. Of course, this does not imply that $\mu(P_5) < \chi_m(P_5)$ and, in fact, $\chi_m(P_5) = \mu(P_5) = 2$ since P_5 is bipartite. While we have seen graphs G for which $\mu(G) = \chi_m(G)$ and graphs G for which $\mu(G) > \chi_m(G)$ (indeed for which $\mu(G)$ is considerably larger than $\chi_m(G)$), we do not know if these are the only possibilities.

Question 9.4.5 ([25]). *Does there exist a graph G for which $\mu(G) < \chi_m(G)$?*

There is a concept closely related to the metric chromatic number of a graph. The *partition dimension* of a connected graph G was introduced in [17, 18] and defined as the minimum positive integer k for which G has a k-coloring such that $\text{code}_\mu(u) \neq \text{code}_\mu(v)$ for each pair u, v of distinct vertices of G. This concept has also been studied in [14, 68, 73] for example. While the partition dimension deals with *vertex-distinguishing* colorings of G (in which every two vertices of G have distinct color-induced labels), a metric coloring is a *neighbor-distinguishing* coloring of G (in which every two *adjacent* vertices of G have distinct color-induced labels). A natural question here is to study the relationship between these two concepts.

Chapter 10
Sigma Colorings

In the preceding three chapters, from a given nonproper vertex coloring of a graph, three proper vertex colorings have been introduced, one involving sets of colors of the original coloring, one involving multisets and one defined in terms of distances. In this chapter, a fourth proper vertex coloring is obtained from the initial vertex coloring (where the colors are positive integers) by addition of colors. In this case, we see that the resulting chromatic number equals one of these three chromatic numbers but with this addition, new concepts arise.

10.1 Sigma Chromatic Number

For a nontrivial connected graph G, let $c : V(G) \to \mathbb{N}$ be a vertex coloring of G where adjacent vertices may be colored the same. For a set S of integers, let $\sigma(S)$ denote the sum of all elements in S. The *color sum* $\sigma(v)$ of v is the sum of the colors of the vertices in $N(v)$, that is, $\sigma(v) = \sigma(M(v))$, where $M(v)$ is the multiset of colors of the neighbors of v (as defined in Chap. 8). If $\sigma(x) \neq \sigma(y)$ for every two adjacent vertices x and y of G, then c is called a *sigma coloring* of G. The minimum number of colors required in a sigma coloring of a graph G is called the *sigma chromatic number* of G and is denoted by $\sigma(G)$. These concepts were introduced and studied in [24].

A graph G with chromatic number 3 is shown in the first diagram in Fig. 10.1 along with a proper coloring of G using the colors 1, 2, 3. Since $\sigma(u) = \sigma(v) = \sigma(y) = 5$, this coloring is not a sigma coloring. However, if we were to interchange the colors 2 and 3 in this diagram, then we obtain a sigma coloring shown in the second diagram of Fig. 10.1.

While not every proper coloring of a graph is a sigma coloring, it is the case that *some* proper coloring of a graph G using $\chi(G)$ colors *is* a sigma coloring.

© The Author 2016
P. Zhang, *A Kaleidoscopic View of Graph Colorings*, SpringerBriefs in Mathematics,
DOI 10.1007/978-3-319-30518-9_10

Fig. 10.1 A non-sigma
coloring and a sigma coloring
of a graph

G:

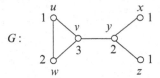 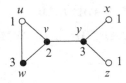

Theorem 10.1.1 ([24]). *For every graph G, $\sigma(G) \leq \chi(G)$.*

Proof. It suffices to assume that G is a nontrivial connected graph. Suppose that $\chi(G) = k$ and $\Delta(G) = \Delta$, and let $d = \Delta + 1$. Let c' be a proper k-coloring of G, using the colors $1, 2, \ldots, k$. So adjacent vertices of G are colored differently by c'. Define a k-coloring c of G by $c(v) = d^{i-1}$ if $c'(v) = i$. We show that c is a sigma coloring of G, which implies that $\sigma(G) \leq k = \chi(G)$.

Let u and v be two adjacent vertices of G, where $c'(u) = s$ and $c'(v) = t$. Then $s \neq t$. Let S be the multiset of colors assigned by c' to the neighbors of u and let T be the multiset of colors assigned by c' to the neighbors of v. Since $s \in T - S$ and $t \in S - T$, it follows that $S \neq T$. Hence, there is a largest integer j with $\max\{s, t\} \leq j \leq k$ such that S and T contain j an unequal number of times. Suppose that S contains j a total of a times and T contains j a total of b times, where say $0 \leq a < b \leq \Delta$.

We claim that $\sigma(u) \neq \sigma(v)$ for the coloring c. Let

$$r = \sum_{i=0}^{k-j-1} a_i d^{k-1-i},$$

where $a_0, a_1, \ldots, a_{k-j-1}$ are nonnegative integers. From the defining property of j, it follows that

$$\sigma(u) = r + ad^{j-1} + p_1(d) \quad \text{and} \quad \sigma(v) = r + bd^{j-1} + p_2(d),$$

where $p_1(d)$ and $p_2(d)$ are polynomials in d having degree at most $j - 2$. Thus

$$\sigma(u) = r + ad^{j-1} + p_1(d) \leq r + ad^{j-1} + \Delta d^{j-2}$$
$$< r + ad^{j-1} + d^{j-1} = r + (a+1)d^{j-1} \leq r + bd^{j-1},$$

while

$$\sigma(v) = r + bd^{j-1} + p_2(d) \geq r + bd^{j-1}.$$

Thus, $\sigma(u) < \sigma(v)$ and so c is a sigma coloring of G. ☐

The following are consequences of the preceding results and observations.

Corollary 10.1.2. *If a nontrivial connected graph G has a proper k-coloring, then G has a sigma k-coloring.*

Corollary 10.1.3. *Let G be a nontrivial connected graph of order n. Then $\sigma(G) = n$ if and only if $G = K_n$.*

Corollary 10.1.4. *If G is a nontrivial connected bipartite graph, then $\sigma(G) \leq 2$. In particular, $\sigma(G) = 1$ if and only if every two adjacent vertices of G have different degrees. Therefore,*

$$\sigma(K_{s,t}) = \begin{cases} 1 & \text{if } s \neq t \\ 2 & \text{if } s = t. \end{cases}$$

10.2 Sigma Colorings Versus Multiset Colorings

Perhaps unexpectedly, the sigma chromatic number equals the multiset chromatic number of every nontrivial connected graph. This fact was established in [25]. To verify this, the following lemma is useful.

Lemma 10.2.1 ([24]). *For integers $k \geq 1$ and $N \geq 1$, let $\mathfrak{A}_k = \{a_1, a_2, \ldots, a_k\}$ be a set of k positive integers such that $a_{i+1} \geq Na_i + 1$ for $1 \leq i \leq k - 1$. Then for every two distinct multisets X and Y of cardinality at most N whose elements belong to $\mathfrak{A}_k$, $\sigma(X) \neq \sigma(Y)$.*

Theorem 10.2.2 ([25]). *For every nontrivial connected graph G, $\chi_m(G) = \sigma(G)$.*

Proof. Since every sigma coloring of G is a multiset coloring of G, it follows that $\chi_m(G) \leq \sigma(G)$. It only remains therefore to show that $\chi_m(G) \geq \sigma(G)$. Suppose that $\chi_m(G) = k$ and $\Delta(G) = \Delta$. Let c be a multiset k-coloring of G using the colors $1, 2, \ldots, k$. Now let $\mathfrak{A}_k = \{a_1, a_2, \ldots, a_k\}$ be a set of k integers, where the elements a_i $(1 \leq i \leq k)$ are defined recursively by (i) $a_1 \geq 1$ and (ii) once a_{i-1} is defined for an integer i with $2 \leq i \leq k$, a_i is an integer such that $a_i \geq \Delta a_{i-1} + 1$. Thus, $a_1 < a_2 < \cdots < a_k$. Define a k-coloring c' of G by

$$c'(v) = a_{c(v)} \text{ for } v \in V(G) \text{ and } 1 \leq i \leq k.$$

We show that c' is a sigma coloring of G. Let x and y be two adjacent vertices of G. Then $M(x) \neq M(y)$. Let S_x be the submultisets of $\mathfrak{A}_k$ obtained from $M(x)$ by replacing each element $i \in M(x)$ by a_i. Similarly, S_y is the multisubset of $\mathfrak{A}_k$ obtained from $M(y)$ by replacing each element $i \in M(x)$ by a_i. Thus, S_x and S_y are two distinct submultisets of $\mathfrak{A}_k$. Since $|S_x| \leq \Delta$ and $|S_y| \leq \Delta$, it follows by Lemma 10.2.1 that $\sigma(S_x) \neq \sigma(S_y)$. $\square$

While $\sigma(G) = \chi_m(G)$ for every graph G, there are major differences between these two colorings. In any multiset coloring of a graph G, it is not important which colors are used; that is, if c is a multiset k-coloring of a graph G, then any k positive integers can be used for the colors. As we saw in Fig. 10.1, this is not the case for a sigma k-coloring of G, however. In order to see some examples of these, we first present the sigma chromatic numbers of regular complete multipartite graphs and complete multipartite graphs, which equal the multiset chromatic numbers of these graphs by Theorem 10.2.2. Recall that for positive integers ℓ and n, the number of n-element submultisets of an ℓ-element set is denoted by $f(\ell, n) = \binom{n+\ell-1}{\ell-1}$.

Theorem 10.2.3. *For positive integers k and n, the sigma chromatic number of the regular complete k-partite graph $K_{k(n)}$ is the unique positive integer ℓ for which*

$$f(\ell - 1, n) < k \leq f(\ell, n).$$

Theorem 10.2.4. *Let $G = K_{k_1(n_1), k_2(n_2),...,k_t(n_t)}$, where $n_1, n_2, \ldots, n_t$ are t distinct positive integers. Then*

$$\sigma(G) = \max\{\sigma(K_{k_i(n_i)}) : 1 \leq i \leq t\}.$$

If $G = K_{10(3)}$, then $\sigma(G) = \chi_m(G) = 3$ by Theorem 10.2.3. Of course, there is a multiset 3-coloring of G using the colors 1, 2, 3. There is, however, no such sigma 3-coloring. In fact, there is no sigma 3-coloring that uses any three of the four colors of the set $\{1, 2, 3, 4\}$. On the other hand, there is a sigma 3-coloring of G using the colors 1, 2, 5.

By Theorem 10.2.3, $\sigma(K_{5(2)}) = 3$ and $\sigma(K_{3(3)}) = 2$. Thus, it follows by Theorem 10.2.4 that $\sigma(K_{5(2),3(3)}) = 3$. In Fig. 10.2, a sigma 3-coloring of the complete 8-partite graph $K_{5(2),3(3)}$ is given using the colors 1, 2, 3, where the color sums of the vertices in each partite set are given as well. The colors assigned to the vertices in this coloring are not interchangeable, however, as every non-identity permutation ϕ of the colors 1, 2, 3 produces a 3-coloring that is not a sigma coloring, as is also shown in Fig. 10.2.

10.3   Sigma Value and Range

If G is a graph with chromatic number k, then, of course, the minimum number of colors that need to be assigned to the vertices of G so that adjacent vertices are colored differently is k. If, as is customary, the colors are positive integers, then the smallest positive integer ℓ such that each color which can be assigned to a vertex of G that is at most ℓ is also k. That is, for every k-chromatic graph G, there is always a coloring of G using the colors $1, 2, \ldots, k$. In other words, the chromatic number $\chi(G)$ can be defined as (1) the minimum number of colors needed to color the vertices of G so that adjacent vertices are colored differently or as (2) the minimum among all largest colors used to color the vertices of G with positive integers so that adjacent vertices are colored differently. As we have seen, such is not the case with sigma colorings. These observations give rise to another concept.

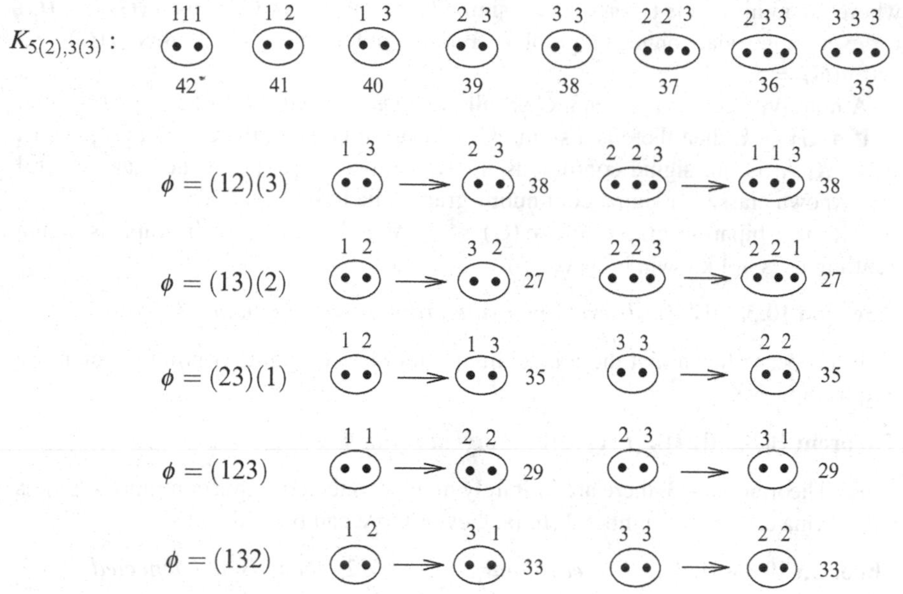

Fig. 10.2 3-Colorings of $K_{5(2),3(3)}$

For a nontrivial connected graph G with $\sigma(G) = k$ and a sigma k-coloring c of G, let

$$\min(c) = \min\{c(v) : v \in V(G)\} \quad \text{and} \quad \max(c) = \max\{c(v) : v \in V(G)\}.$$

That is, $\min(c)$ is the smallest color assigned by c to a vertex of G and $\max(c)$ is the largest such color. Certainly,

$$\max(c) \geq \min(c) + (k - 1) \geq k$$

for every sigma k-coloring of a graph G with $\sigma(G) = k$. In fact, there is always some sigma k-coloring c of G such that $\min(c) = 1$.

Proposition 10.3.1 ([24]). *Suppose that G is a nontrivial connected graph with $\sigma(G) = k$. Then there is a sigma k-coloring of G that assigns some vertex of G the color 1.*

A question of interest concerns the minimum value of $\max(c)$ over all sigma k-colorings c of G for which $\sigma(G) = k$. This minimum value is called the *sigma value* $\upsilon(G)$ of G. That is, for a connected graph G with $\sigma(G) = k$,

$$\upsilon(G) = \min\{\max(c)\},$$

where the minimum is taken over all sigma k-colorings c of G. Thus, $\nu(G) \geq \sigma(G)$ for every nontrivial connected graph G. For example, if $G = K_{10(3)}$, then $\sigma(G) = 3$ and $\nu(G) = 5$.

A nontrivial connected graph G is called *sigma continuous* if $\nu(G) = \sigma(G)$, that is, if $\sigma(G) = k$, then there is a sigma k-coloring of G using the colors in $[k]$. Thus, $G = K_{10(3)}$ is not sigma continuous. It was shown in [24] that there are several well-known classes of sigma continuous graphs, including cycles.

If G is a bipartite graph, then $\sigma(G) \leq 2$. Whether every such graph is sigma continuous is not known, however.

Question 10.3.2 ([24]). *Is every bipartite graph sigma continuous?*

It has been shown that there is an important class of sigma continuous bipartite graphs, however.

Theorem 10.3.3 ([24]). *Every tree is sigma continuous.*

By Theorem 10.3.3, there are infinitely many connected sigma continuous graphs with sigma chromatic number 2. In fact, even more can be said.

Theorem 10.3.4 ([24]). *For each integer $k \geq 2$, there is a connected sigma continuous graph with sigma chromatic number k.*

As a consequence of Theorem 10.2.4, for integers n_1 and n_2 where $1 \leq n_1 < n_2$ and $k_i = n_i + 1$ $(i = 1, 2)$, if $G = K_{k_1(n_1), k_2(n_2)}$, then $\sigma(G) = 2$. The admissible sigma 2-colorings of G were established in [24].

Theorem 10.3.5 ([24]). *For integers n_1 and n_2 with $1 \leq n_1 < n_2$ and $k_i = n_i + 1$ $(i = 1, 2)$, let $G = K_{k_1(n_1), k_2(n_2)}$. For positive integers a and b, there exists a sigma 2-coloring of G using the colors a and $a + b$ if and only if*

$$(a + b)n_1 < an_2 \text{ or } a(n_2 - n_1) \not\equiv 0 \pmod{b}.$$

By Theorem 10.3.5, there exist connected graphs with sigma chromatic number 2 that is not sigma continuous. In fact, more can be said.

Theorem 10.3.6 ([24]). *For each integer $k \geq 2$, there is a connected graph with sigma chromatic number k that is not sigma continuous.*

Another parameter of interest was introduced in [24]. For a sigma coloring c of G, the *sigma range* $\rho(G)$ of G is defined by

$$\rho(G) = \min\{\max(c)\}$$

over all sigma colorings c of G. Hence, the sigma range of G is the smallest positive integer k for which there exists a sigma coloring of G using colors from the set $[k]$, while the sigma value of G is the smallest positive integer k for which there exists a sigma coloring of G using $\sigma(G)$ colors from the set $[k]$. Therefore, for every graph G,

$$\sigma(G) \leq \rho(G) \leq \nu(G).$$

As an example, consider $G = K_{10(3)}$ again. We saw that $\sigma(G) = 3$ and $\nu(G) = 5$. We now show that $\rho(G) = 4$. Since there is no sigma coloring using the colors 1, 2, 3, it follows that $\rho(G) \geq 4$. To show that $\rho(G) \leq 4$, let $V_1, V_2, \ldots, V_{10}$ be the partite sets of G and let $A_1, A_2, \ldots, A_{10}$ be the following 3-element submultisets of $\{1, 2, 3, 4\}$:

$$\{1, 1, 1\}, \ \{1, 1, 2\}, \ \{1, 1, 3\}, \ \{1, 1, 4\}, \ \{1, 2, 4\},$$
$$\{1, 3, 4\}, \ \{1, 4, 4\}, \ \{2, 4, 4\}, \ \{3, 4, 4\}, \ \{4, 4, 4\}.$$

Since $\sigma(A_i) \neq \sigma(A_j)$ for $1 \leq i < j \leq 10$, the 4-coloring of G that assigns the three colors in A_i to the three vertices in V_i for each i with $1 \leq i \leq 10$ is a sigma coloring of G using the colors $1, 2, 3, 4$. Thus, $\rho(G) = 4$. Therefore, if $G = K_{10(3)}$, then $\sigma(G) = 3$, $\rho(G) = 4$ and $\nu(G) = 5$.

It was shown in [24] that $\nu(G) - \rho(G)$ can be arbitrarily large for some connected graphs G. Such is also the case for $\rho(G) - \sigma(G)$. The following problem appears in [24].

Question 10.3.7 ([24]). *Which ordered triples of positive integers can be realized as $(\sigma(G), \rho(G), \nu(G))$ for some graph G?*

10.4   Four Colorings Problems

We have seen that for each of the four neighbor-distinguishing vertex colorings

set colorings, metric colorings, multiset colorings, sigma colorings,

the number of colors required to color the vertices of a graph need not exceed the chromatic number of the graph. Thus, we have the following.

Four Four Color Theorems Let G be a nontrivial connected graph.

(1) If G is planar, then $\chi(G) \leq 4$.
(2) If G is planar, then $\chi_s(G) \leq 4$.
(3) If G is planar, then $\mu(G) \leq 4$.
(4) If G is planar, then $\chi_m(G) \leq 4$.

From what we saw in Sect. 10.2, statement (4) can be replaced by

(4′) If G is planar, then $\sigma(G) \leq 4$.

Of course, statements (2), (3) and (4) (and (4′)) are all corollaries of statement (1) (*the* Four Color Theorem). We therefore close with the following.

Question 10.4.1. *Does there exist a proof of any of the statements (2), (3), (4) or (4′), that does not use the original Four Color Theorem and that is not computer-aided?*

Chapter 11
Modular Colorings

Historically, a number of problems and puzzles have been introduced that initially appeared to have no connection to graph colorings but, upon further analysis, suggested graph coloring problems. In this chapter, we discuss two combinatorial problems and two graph coloring problems inspired by these problems.

11.1 A Checkerboard Problem

Suppose that the squares of an $m \times n$ checkerboard (m rows and n columns), where $1 \le m \le n$ and $n \ge 2$, are alternately colored black and red. Figure 11.1 shows a 5×7 checkerboard where a shaded square represents a black square. Two squares are said to be *neighboring* if they belong to the same row or to the same column and there is no square between them. Thus, every two neighboring squares are of different colors. A combinatorial problem was introduced by Gary Chartrand in 2010 and the following conjecture was stated [56].

The Checkerboard Conjecture *It is possible to place coins on some of the squares of an $m \times n$ checkerboard (at most one coin per square) such that for every two squares of the same color the numbers of coins on neighboring squares are of the same parity, while for every two squares of different colors the numbers of coins on neighboring squares are of opposite parity.*

Figure 11.2 shows a placement of six coins on a 5×7 checkerboard such that the number of coins on neighboring squares of every red square is even and the number of coins on neighboring squares of every black square is odd. Thus, for every two squares of different colors, the numbers of coins on neighboring squares are of opposite parity. Consequently, the Checkerboard Conjecture is true for a 5×7 checkerboard. Observe that all six coins on the 5×7 checkerboard of Fig. 11.2 are

© The Author 2016
P. Zhang, *A Kaleidoscopic View of Graph Colorings*, SpringerBriefs in Mathematics,
DOI 10.1007/978-3-319-30518-9_11

Fig. 11.1 A 5 × 7
checkerboard

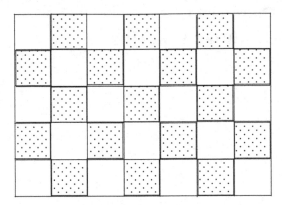

Fig. 11.2 A coin placement
on the 5 × 7 checkerboard

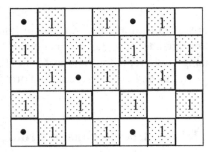

placed only on red squares. Thus, the number of coins on neighboring squares of
every red square is 0 and is therefore even, while the number of coins on neighboring
squares of each black square is 1 and this is shown in Fig. 11.1 as well.

In [56] it was shown that this problem can be placed in a graph theory setting. For
example, let G be the graph whose vertices are the squares of the checkerboard and
where two vertices of G are adjacent if the corresponding squares are neighboring.
Then G is the *grid* (bipartite graph) $P_m \square P_n$ (or $P_m \times P_n$) which is the Cartesian
product of the paths P_m and P_n. This suggests a function (coloring) c on $G =
P_m \square P_n$, where $c : V(G) \to \mathbb{Z}_2$ such that

$$c(v) = \begin{cases} 0 & \text{if } v \text{ corresponds to a square with no coin} \\ 1 & \text{if } v \text{ corresponds to a square containing a coin.} \end{cases}$$

This induces another coloring $\sigma : V(G) \to \mathbb{Z}_2$ defined by

$$\sigma(v) = \sum_{u \in N(v)} c(u) \text{ in } \mathbb{Z}_2, \tag{11.1}$$

where $N(v)$ is the neighborhood of a vertex v and addition is performed in $\mathbb{Z}_2$. If σ is
a proper coloring, then the checkerboard problem has a solution on the checkerboard

represented by G. With the aid of graph colorings described above, it was shown in [57] that the Checkerboard Conjecture is true for a checkerboard of any size.

The Checkerboard Theorem *For every pair m, n of positive integers, it is possible to place coins on some of the squares of an $m \times n$ checkerboard (at most one coin per square) such that for every two squares of the same color the numbers of coins on neighboring squares are of the same parity, while for every two squares of different colors the numbers of coins on neighboring squares are of opposite parity.*

If $\mathbb{Z}_2$ is replaced by $\mathbb{Z}_k$ for an integer $k \geq 2$, then this checkerboard problem gave rise to a new coloring in [56], which we discuss next.

11.2 Modular Colorings

In 2010, a proper vertex coloring was introduced in [56] for the purpose of finding solutions to the checkerboard problem described in Sect. 11.1. For a nontrivial connected graph G, let $c : V(G) \to \mathbb{Z}_k$ ($k \geq 2$) be a vertex coloring of G where adjacent vertices may be colored the same. The *color sum* $\sigma(v)$ of a vertex v of G is defined as

$$\sigma(v) = \sum_{u \in N(v)} c(u) \text{ in } \mathbb{Z}_k, \qquad (11.2)$$

where the addition is performed in $\mathbb{Z}_k$. Thus, the vertex coloring c induces another vertex coloring $\sigma : V(G) \to \mathbb{Z}_k$ of G. If $\sigma(x) \neq \sigma(y)$ in $\mathbb{Z}_k$ for every two adjacent vertices x and y of G, then the coloring c is called a *modular k-coloring* of G. The minimum k for which G has a modular k-coloring is called the *modular chromatic number* of G and is denoted by $mc(G)$. Modular colorings in graphs have been studied in [4, 38, 55–57]. To illustrate the concepts introduced above, Fig. 11.3 shows a modular 3-coloring of a bipartite graph G (where the color of a vertex is placed within the vertex) together with the color sum $\sigma(v)$ for each vertex v of G (where the color sum of a vertex is placed next to the vertex). In fact, $mc(G) = 3$ for this graph G.

The Checkerboard Theorem can, therefore, be stated in terms of graphs and modular colorings as follows.

Fig. 11.3 A bipartite graph G with $mc(G) = 3$

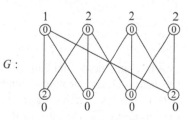

Theorem 11.2.1 ([57]). *For every two positive integers m and n with mn $\geq$ 2,*

$$mc(P_m \,\square\, P_n) = 2.$$

Modular colorings are related to another neighbor-distinguishing vertex color-ings of a graph, namely *sigma colorings*, which were described in Chap. 10. For sigma colorings, a given coloring c is a function $c : V(G) \to \mathbb{N}$ and the color sum $\sigma(v)$ of a vertex v has the same formula as defined in (11.2) except that addition is performed in $\mathbb{N}$ rather than $\mathbb{Z}_k$ for some integer $k \geq 2$. Suppose that $V(G) = \{v_1, v_2, \ldots, v_n\}$. Define a coloring c of G by $c(v_i) = 2^{i-1}$ for $1 \leq i \leq n$. Let $k = \sum_{i=1}^{n-1} 2^i = 2(2^{n-1} - 1)$. Considering $c : V(G) \to \mathbb{Z}_k$, it follows that $1 \leq \sigma(v_i) \leq k$ for all $i (1 \leq i \leq n)$ and $\sigma(v_i) \neq \sigma(v_j)$ whenever v_i and v_j are adjacent. Hence, c is a modular k-coloring of G. Therefore, every nontrivial connected graph G has a modular coloring and so the modular chromatic number of G exists. The following two observations will also be useful to us.

Observation 11.2.2 ([56]). *If u and v are two adjacent vertices in a graph G such that $N(u) - \{v\} = N(v) - \{u\}$, then $c(u) \neq c(v)$ for every modular coloring c of G.*

Observation 11.2.3 ([56]). *If H is a complete subgraph of order k in a graph G, then*

$$mc(G) \geq k.$$

Observation 11.2.3 can be restated to say that $mc(G) \geq \omega(G)$, where $\omega(G)$ denotes the clique number of G. In fact, there is an observation which is even stronger than that given in Observation 11.2.3.

Observation 11.2.4 ([56]). *For every nontrivial connected graph G,*

$$mc(G) \geq \chi(G).$$

As the graph G of Fig. 11.3 shows, the inequality in Observation 11.2.4 can be strict. Figure 11.4 shows a modular 2-coloring of C_8 and modular 3-colorings of C_9, C_{10} and C_{11}.

It turns out that the modular colorings of the four cycles in Fig. 11.4 illustrate the modular chromatic numbers for all cycles.

Proposition 11.2.5 ([56]). *For each integer n $\geq$ 3,*

$$mc(C_n) = \begin{cases} 2 & \text{if } n \equiv 0 \pmod{4} \\ 3 & \text{otherwise.} \end{cases}$$

While each nontrivial path is a tree with modular chromatic number 2, not every tree has modular chromatic number 2. We show for the tree T in Fig. 11.5 that $mc(T) = 3$. Assume, to the contrary, that $mc(T) = 2$. Then there exists a modular 2-coloring c of T. Because of the symmetry of the structure of T, the color sums

Fig. 11.4 Modular colorings
of C_n for $8 \leq n \leq 11$

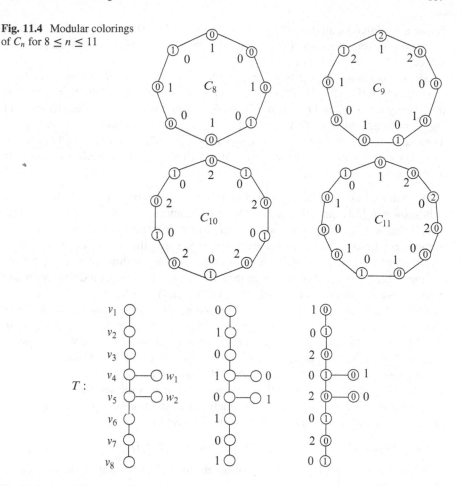

Fig. 11.5 A tree T with $mc(T) = 3$

of the vertices of T are those shown in the second diagram of Fig. 11.5. Since
$\sigma(w_2) = \sigma(v_8) = 1$, it follows that $c(v_5) = c(v_7) = 1$. This, however, contradicts
the fact that $\sigma(v_6) = 1$. Hence, $mc(T) \neq 2$ and so $mc(T) = 3$. A modular
3-coloring of T is also shown in Fig. 11.5.

In fact, every nontrivial tree has modular chromatic number 2 or 3.

Theorem 11.2.6 ([56]). *If T is a nontrivial tree, then* $mc(T) = 2$ *or* $mc(T) = 3$.

Proof. Since $mc(T) \geq \chi(T) = 2$ for every nontrivial tree T, it remains to show that
there is a modular 3-coloring of T. Since $mc(T) = 2$ if T is a nontrivial star, suppose
that $\mathrm{diam}(T) \geq 3$. Let $x \in V(T)$ be an end-vertex with $N(x) = \{y\}$. For each i with
$0 \leq i \leq e(x)$, where $e(x)$ is the eccentricity of x, let $V_i = \{v \in V(T) : d(v,x) = i\}$.
Thus, $V_0 = \{x\}$, $V_1 = \{y\}$ and $\{V_0, V_1, \ldots, V_{e(x)}\}$ is a partition of $V(T)$. We now

define a modular 3-coloring c of T. First, define $c(v) = 0$ if $v \in V_i$ where $0 \leq i \leq$ $e(x)$ and i is even. Also, let $c(y) = 1$. Then $\sigma(x) = 1$ and $\sigma(u) = 0$ for each $u \in V_j$ where $1 \leq j \leq e(x)$ and j is odd.

Since T is not a star, $e(x) \geq 3$. Suppose that $c(v)$ has been defined for all vertices $v \in V_i$, where $0 \leq i \leq 2k < e(x)$ for some positive integer k, such that $c(v) = 0$ if and only if $v \in V_i$ and i is even. If $w \in V_{2k}$ is not an end-vertex, then let $N(w) \cap$ $V_{2k-1} = \{w_0\}$ and $N(w) \cap V_{2k+1} = \{w_1, w_2, \ldots, w_{d-1}\}$, where $d = \deg w \geq 2$. Then define $c(w_i) = 1$ for $1 \leq i \leq d - 1$ if $\sum_{i=0}^{d-1} c(w_i) \not\equiv 0 \pmod 3$. Otherwise, let $c(w_1) = 2$ and $c(w_i) = 1$ for $2 \leq i \leq d - 1$ (if $d \geq 3$). This results in a modular 3-coloring of T and so $\mathrm{mc}(T) \leq 3$.   □

A nontrivial tree T is of *type one* if $\mathrm{mc}(T) = 2$ and is of *type two* if $\mathrm{mc}(T) = 3$. It is shown in [55] that all nontrivial trees of diameter at most 6 are of type one. Recall that a *caterpillar* is a tree of order 3 or more, the removal of whose end-vertices produces a path. A characterization of caterpillars that are of type two was presented in [55]. An efficient algorithm has been established to compute the modular chromatic number of a given tree in [38]. Furthermore, modular chromatic numbers are determined for several classes of graphs in [56].

Of course, every nontrivial tree is a bipartite graph. Furthermore, the modular chromatic number of every nontrivial bipartite graph is at least 2. We saw in Fig. 11.3 that a bipartite graph that is not a tree may also have modular chromatic number 3. Whether a bipartite graph G can have a larger modular chromatic number is not known, but it can never exceed $\Delta(G)$ by more than 1. If G is a bipartite graph with $\Delta(G) = \Delta$, then we assign 0 to each vertex in one partite set of G and 1 to each vertex in the other partite set of G, producing a modular coloring using the colors on $\mathbb{Z}_{1+\Delta}$. Therefore, we have the following observation.

Observation 11.2.7. *If G is a bipartite graph, then $\mathrm{mc}(G) \leq 1 + \Delta(G)$.*

There are some conditions that are sufficient for a bipartite graph to have modular chromatic number 2.

Lemma 11.2.8. *Let G be a bipartite graph. If G contains a vertex that is adjacent to all vertices in a partite set, then $\mathrm{mc}(G) = 2$.*

Proof. Suppose that the partite sets of G are V_1 and V_2 and $v \in V_1$ is adjacent to every vertex in V_2. Then the coloring $c : V(G) \to \mathbb{Z}_2$ defined by $c(v) = 1$ and $c(x) = 0$ for all $x \in V(G) - \{v\}$ is a modular 2-coloring and so $\mathrm{mc}(G) = 2$.   □

Lemma 11.2.9. *If G is a bipartite graph such that one of its partite sets consists only of odd vertices, then $\mathrm{mc}(G) = 2$.*

Proof. Suppose that the partite sets of G are V_1 and V_2 and every vertex in V_1 has odd degree. Then the coloring $c : V(G) \to \mathbb{Z}_2$ defined by $c(v) = 0$ if and only if $v \in V_1$ is a modular 2-coloring and so $\mathrm{mc}(G) = 2$.   □

Proposition 11.2.10. *If G is a bipartite graph the degrees of whose vertices are of the same parity, then $\mathrm{mc}(G \square K_2) = 2$.*

Proof. Since G is a bipartite graph, $G \square K_2$ is also bipartite. If all vertices of G are even, then all vertices of $G \square K_2$ are odd and so $mc(G \square K_2) = 2$ by Lemma 11.2.9. Thus, we may assume that all vertices of G are odd. In this case, each vertex of $G \square K_2$ is even. Suppose that G is of order n and let G_1 and G_2 be two copies of G in $G \square K_2$, where $V(G_1) = \{u_1, u_2, \ldots, u_n\}$, $V(G_2) = \{w_1, w_2, \ldots, w_n\}$ and $u_i w_i \in E(G)$ for $1 \leq i \leq n$. Consider the coloring $c : V(G) \to \mathbb{Z}_2$ such that (i) $c(u) = 0$ for every $u \in V(G_1)$ and (ii) $c : V(G_2) \to \mathbb{Z}_2$ is a proper coloring of G_2. Since $\sigma(u_i) = c(w_i)$ and $\sigma(w_i) = c(w_i) + 1$ for $1 \leq i \leq n$, it follows that $\sigma(u_i) \neq \sigma(w_i)$. Furthermore, if u_i and u_j are adjacent, then w_i and w_j are also adjacent and so $c(w_i) \neq c(w_j)$, implying that $\sigma(u_i) \neq \sigma(u_j)$. Similarly, if w_i and w_j are adjacent, then $\sigma(w_i) \neq \sigma(w_j)$. Therefore, c is a modular 2-coloring of $G \square K_2$. $\square$

A well-known class of bipartite graphs are the n-cubes. By Proposition 11.2.10, all of these graphs have modular chromatic number 2.

Corollary 11.2.11. *For every positive integer n, $mc(Q_n) = 2$.*

We saw in Proposition 11.2.5 that $mc(C_n) = 3$ if $n \geq 6$ and $n \equiv 2 \pmod 4$. Hence, the bound given in Observation 11.2.7 is sharp when $\Delta(G) = 2$. Whether this bound is sharp when $\Delta(G) \geq 3$ is not known, as was noted earlier. Of course, $\chi(C_n) = 2$ when $n \equiv 2 \pmod 4$ and so $mc(C_n) = \chi(C_n) + 1$. This brings up other questions.

Question 11.2.12. *Is there a graph G such that $mc(G) \geq \chi(G) + 2$?*

Question 11.2.13. *Is there a graph G such that $\omega(G) < \chi(G) < mc(G)$?*

The upper bound for $mc(G)$ stated in Observation 11.2.7 can be extended to graphs whose chromatic number is at least 3.

Theorem 11.2.14. *If G is a k-chromatic graph $(k \geq 2)$ with maximum degree Δ, then*

$$mc(G) \leq \Delta(\Delta + 1)^{k-2} + 1.$$

Among other questions on this topic are the following (see [4]).

Question 11.2.15. *Is there a constant C such that $mc(G) \leq C$ for every bipartite graph G?*

Question 11.2.16. *Does there exist a planar graph whose modular chromatic number is 5?*

If the answer to Question 11.2.16 is no (and if we can verify this), then there is a new Four Color Theorem for which the classic Four Color Theorem is a corollary.

11.3   A Lights Out Problem

Another recreational problem concerns the electronic game of "Lights Out" consisting of a cube, each of whose six faces contains nine squares in three rows and three columns. Thus, there are 54 squares in all. The first diagram in Fig. 11.6 shows the "front" of the cube as well as the faces on the top, bottom, left and right. The back of the cube is not shown. A button is placed on each square of a Lights Out cube containing a light which is either on or off. When a button is pushed, the light on that square changes from on to off or from off to on. Moreover, not only is the light on that square reversed when its button is pushed but the lights on its four neighboring squares (top, bottom, left, right) are reversed as well. The four neighboring squares of the middle square of a face lie on the same face as the middle square. Only three neighboring squares of a "side square" (top middle, bottom middle, left middle, right middle) lie on the same face of such a square, with the remaining neighboring square lying on an adjacent face as the middle square. For example, if all 54 lights are on initially and the button on the top middle square on the front face is pushed, then this light goes off as well as the lights on its four neighboring squares (see the second diagram in Fig. 11.6). Only two neighboring squares of a "corner" square lie on the same face as that square; the other two neighboring squares lie on two other faces.

One goal of the game "Lights Out" is to begin with such a cube where all lights are on and to push a set of buttons so that, at the end, all lights are out. Two observations are immediate: (1) No button needs to be pushed more than once. (2) The order in which the buttons are pushed is immaterial.

This game has a setting in graph theory. Let each square be a vertex and join each vertex to the vertices corresponding to its neighboring squares. This results in a 4-regular graph G of order 54. The goal is to locate a collection S of vertices of G, which correspond to the buttons to be pushed, such that every vertex of G is in the closed neighborhood of an odd number of vertices of S. This says that each vertex v

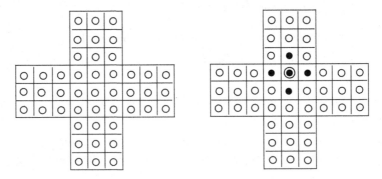

Fig. 11.6 Lights Out Game

that corresponds to a lit square will have its light reversed an odd number of times, resulting in the light being turned out.

The Lights Out Game can in fact be played on any connected graph G on which there is a light at each vertex of G. The game has a solution for the graph G if all lights are on initially and if there exists a collection S of vertices which, when the button on each vertex of S is pushed, all lights of G will be out. This problem has another interpretation. A vertex v of a graph G *dominates* a vertex u if u belong to the closed neighborhood $N[v]$ of v (consisting of v and the vertices in the (open) neighborhood $N(v)$ of v). The Lights Out Game has a solution for a given graph G if and only if G contains a set S of vertices such that every vertex of G is dominated by an odd number of vertices of S. In [71] Sutner showed that every graph has this property and so the Lights Out Game is solvable on every graph.

As discussed in [26], the Lights Out Game is also equivalent to beginning with a connected graph G where every vertex of G is initially assigned the color 1 in $\mathbb{Z}_2$ (corresponding to its light being on) and finding a set S of vertices of G and a coloring $c : V(G) \to \mathbb{Z}_2$ such that

$$c(v) = \begin{cases} 1 & \text{if } v \in S \\ 0 & \text{if } v \notin S. \end{cases}$$

A new coloring $\sigma' : V(G) \to \mathbb{Z}_2$ induced by c is defined by

$$\sigma'(v) = 1 + \sum_{u \in N[v]} c(u) \text{ in } \mathbb{Z}_2. \tag{11.3}$$

The goal of the Lights Out Game is therefore to have $\sigma'(v) = 0$ for all $v \in V(G)$. The Lights Out Game suggests a coloring problem introduced in [26], which we will discuss in Sect. 11.4.

11.4 Closed Modular Colorings

Modular colorings and the Lights Out Game suggest other coloring problems. For a positive integer k and a connected graph G, let $c : V(G) \to \mathbb{Z}_k$ be a vertex coloring where adjacent vertices may be assigned the same color. The *closed color sum* $\overline{\sigma}(v)$ of v is defined by

$$\overline{\sigma}(v) = \sum_{u \in N[v]} c(u) \text{ in } \mathbb{Z}_k \tag{11.4}$$

where the addition is performed in $\mathbb{Z}_k$. Thus, c induces another vertex coloring $\overline{\sigma} : V(G) \to \mathbb{Z}_k$ of G. If u and v are adjacent vertices of a graph G such that $N[u] = N[v]$, then $\overline{\sigma}(u) = \overline{\sigma}(v)$ for every vertex coloring c of G. Therefore, the coloring $\overline{\sigma}$

cannot be neighbor-distinguishing in general. For this reason, additional restrictions are needed.

Two vertices u and v in a connected graph G are *twins* if u and v have the same neighbors in $V(G) - \{u, v\}$. If u and v are adjacent, they are referred to as *true twins*; while if u and v are nonadjacent, they are *false twins*. If u and v are adjacent vertices of a graph G such that $N[u] = N[v]$, that is, if u and v are true twins, then $\overline{\sigma}(u) = \overline{\sigma}(v)$ for every vertex coloring c of G. Define a coloring $c : V(G) \to \mathbb{Z}_k$ to be a *closed modular k-coloring* if $\overline{\sigma}(u) \neq \overline{\sigma}(v)$ in $\mathbb{Z}_k$ for all pairs u, v of adjacent vertices for which $N[u] \neq N[v]$ in G (or u and v are true twins in G). A vertex coloring c is a *closed modular coloring* of G if c is a closed modular k-coloring of G for some positive integer k. That is, in a closed modular coloring c of a graph, $\overline{\sigma}(u) = \overline{\sigma}(v)$ if u and v are true twins, $\overline{\sigma}(u) \neq \overline{\sigma}(v)$ if u and v are adjacent vertices that are not true twins and no condition is placed on $\overline{\sigma}(u)$ and $\overline{\sigma}(v)$ otherwise. The minimum k for which G has a closed modular k-coloring is called the *closed modular chromatic number* of G and is denoted by $\overline{mc}(G)$. Since every two vertices in a nontrivial complete graph K_n are true twins, it follows that $\overline{mc}(K_n) = 1$ for all $n \geq 2$. In fact, for each integer $n \geq 2$, the complete graph K_n of order n is the only connected graph G of order n for which $\overline{mc}(G) = 1$.

Proposition 11.4.1. *Let G be a nontrivial connected graph. Then $\overline{mc}(G) = 1$ if and only if $G = K_n$ for some $n \geq 2$.*

Proof. We saw that $\overline{mc}(K_n) = 1$. Thus, it remains to verify the converse. Let G be a connected graph of order n that is not complete. Then $n \geq 3$ and G contains a $u - w$ geodesic (u, v, w) of length 2 for some pair u, w of vertices. Since $w \in N[v] - N[u]$, it follows that $N[u] \neq N[v]$. Since the closed color sums of u and v in every closed modular coloring of G are different, $\overline{mc}(G) \geq 2$.   □

To illustrate these concepts, Fig. 11.7 shows a closed modular 3-coloring of a bipartite graph G (where the color of a vertex is placed within the vertex) together with the color sum $\overline{\sigma}(v)$ for each vertex v of G (where the color sum of a vertex is placed next to the vertex). In fact, $\overline{mc}(G) = 3$ for this graph G. This example illustrates the following useful result.

Proposition 11.4.2. *Let c be a closed modular coloring of a connected graph G and let u and v be false twins in G. Then $c(u) = c(v)$ if and only if $\overline{\sigma}(u) = \overline{\sigma}(v)$.*

Fig. 11.7 A bipartite graph
G with $\overline{mc}(G) = 3$

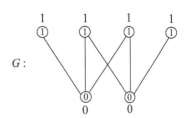

These concepts were introduced and studied in [26] and studied further in [4, 60–63]. Next, we present some results obtained on this topic (see [60]).

Proposition 11.4.3. *If G is a nontrivial connected graph, then $\overline{mc}(G)$ exists. Furthermore, if G does not contain a pair of true twins, then $\overline{mc}(G) \geq \chi(G)$.*

Proof. Let $V(G) = \{v_1, v_2, \ldots, v_n\}$ where $n \geq 2$ and $k = \sum_{i=0}^{n-1} 2^i = 2^n - 1$. Define a coloring $c : V(G) \to \mathbb{Z}_k$ of G by $c(v_i) = 2^{i-1}$ for $1 \leq i \leq n$. Then $1 \leq \overline{\sigma}(v_i) \leq k$ for all $i (1 \leq i \leq n)$. If v_i and v_j are adjacent vertices that are not true twins, then $N[v_i] \neq N[v_j]$ and $\overline{\sigma}(v_i) \neq \overline{\sigma}(v_j)$. Hence, c is a closed modular k-coloring of G and so $\overline{mc}(G)$ exists. Since this is a proper vertex coloring of G using at most $\overline{mc}(G)$ colors, it follows that $\chi(G) \leq \overline{mc}(G)$.   □

For an edge uv of a graph G, the graph G/uv is obtained from G by *contracting the edge uv*. Thus, $V(G/uv)$ is obtained by identifying u and v in $V(G)$. If we denote the vertex $u = v$ in G/uv by w, then $V(G/uv) = (V(G) \cup \{w\}) - \{u, v\}$ and the edge set of G/uv is

$$E(G/uv) = \{xy : xy \in E(G), x, y \in V(G) - \{u, v\}\} \cup$$

$$\{wx : ux \in E(G) \text{ or } vx \in E(G), x \in V(G) - \{u, v\}\}.$$

The graph G/uv is referred to as an *elementary contraction* of G.

Theorem 11.4.4 ([60]). *Let u and v be true twins of a nontrivial connected graph G. Then G has a closed modular k-coloring if and only if G/uv has a closed modular k-coloring.*

Proof. Suppose that w is the vertex of G/uv obtained by identifying u and v in G. Let $c_0 : V(G) \to \mathbb{Z}_k$ be a closed modular k-coloring of G. Since u and v are true twins of G, it follows that

$$\overline{\sigma}_0(u) = \overline{\sigma}_0(v) = c_0(u) + c_0(v) + \sum_{x \in N(u) - \{v\}} c_0(x).$$

Define the coloring $c : V(G/uv) \to \mathbb{Z}_k$ by $c(x) = c_0(x)$ if $x \neq w$ and $c(w) = c_0(u) + c_0(v)$. Then

$$\overline{\sigma}(w) = c(w) + \sum_{x \in N_{G/uv}(w)} c_0(x) = \overline{\sigma}_0(u) = \overline{\sigma}_0(v).$$

Furthermore, if $z \neq w$ in G/uv, then $\overline{\sigma}(z) = \overline{\sigma}_0(z)$. Thus, c is a closed modular k-coloring of G/uv.

For the converse, let $c_0 : V(G/uv) \to \mathbb{Z}_k$ be a closed modular k-coloring of G/uv. Define the coloring $c : V(G) \to \mathbb{Z}_k$ by $c(x) = c_0(x)$ if $x \notin \{u, v\}$, $c(u) = c_0(w)$ and $c(v) = 0$. Then

$$\overline{\sigma}(u) = \overline{\sigma}(v) = c(u) + c(v) + \sum_{x \in N(u) - \{v\}} c(x)$$

$$= c_0(w) + \sum_{x \in N(u) - \{v\}} c_0(x) = \overline{\sigma}_0(w).$$

Furthermore, if $x \notin \{u, v\}$ in G, then $\overline{\sigma}(x) = \overline{\sigma}_0(x)$. Thus, c is a closed modular k-coloring of G. $\square$

For a nontrivial connected graph G, define the *true twins closure* $TC(G)$ of G as the graph obtained from G by a sequence of elementary contractions of pairs of true twins in G until no such pair remains. In particular, if G contains no true twins, then $TC(G) = G$. A *minor* of a graph G is either isomorphic to G or can be obtained from G by a succession of edge contractions, edge deletions or vertex deletions in any order. Thus, $TC(G)$ is a minor of G

The following then is a consequence of Theorem 11.4.4.

Corollary 11.4.5. *For a nontrivial connected graph G, $\overline{mc}(G) = \overline{mc}(TC(G))$.*

By Proposition 11.4.3, if G is a nontrivial connected graph that contains no true twins, then $\overline{mc}(G) \geq \chi(G)$. On the other hand, if G contains true twins, then it is possible that $\overline{mc}(G) < \chi(G)$. In fact, more can be said.

Theorem 11.4.6. *For each pair a, b of positive integers with $a \leq b$ and $b \geq 2$, there is a connected graph G such that $\overline{mc}(G) = a$ and $\chi(G) = b$.*

Proof. First, suppose that $a = b \geq 2$. Let $G = K_{2,2,\dots,2}$ be the regular complete a-partite graph with partite sets $U_1, U_2, \dots, U_a$ such that $|U_i| = 2$ for $1 \leq i \leq a$. Then $\chi(G) = a$. Since G contains no true twins, it follows by Proposition 11.4.3 that $\overline{mc}(G) \geq a$. Next, define the coloring $c : V(G) \to \mathbb{Z}_a$ by $c(x) = i - 1$ if $x \in U_i$ for $1 \leq i \leq a$. Thus, if $u_i \in U_i$ $(1 \leq i \leq a)$, then

$$\overline{\sigma}(u_i) = \left(\sum_{j=1}^{a} [2(j-1)] \right) - 2(i-1) + (i-1)$$

$$= 2\binom{a}{2} - (i-1) = -(i-1) \text{ in } \mathbb{Z}_a.$$

Hence, $\overline{\sigma}(u_i) \neq \overline{\sigma}(u_j)$ if $u_i \in U_i$ and $u_j \in U_j$ where $i \neq j$ and $1 \leq i, j \leq a$. Thus, $\overline{\sigma}$ is a proper vertex coloring of G. Since c is a closed modular a-coloring of G, it follows that $\overline{mc}(G) = a$.

Next, assume that $a < b$. If $a = 1$, then let $G = K_b$ and the result follows by Proposition 11.4.1. For $a \geq 2$, let G be the graph obtained from the regular complete a-partite graph $K_{2,2,\ldots,2}$ by replacing a vertex v of $K_{2,2,\ldots,2}$ by the complete graph K_{b-a+1} and joining each vertex of K_{b-a+1} to all the neighbors of v in $K_{2,2,\ldots,2}$. Thus, G contains K_b as a subgraph. Since $TC(G) = K_{2,2,\ldots,2}$, it follows by Corollary 11.4.5 that $\overline{mc}(G) = \overline{mc}(K_{2,2,\ldots,2}) = a$. Furthermore, $\chi(G) = \chi(K_b) = b$. $\square$

By Corollary 11.4.5, it suffices to consider nontrivial connected graphs containing no true twins. Closed modular chromatic numbers are determined for several classes of regular graphs (see [26]). For every connected graph G without true twins that we have encountered, $\overline{mc}(G) \leq 2\chi(G) - 1$. Thus, the following is the primary open question on this topic.

Question 11.4.7. *Let G be a connected graph G of order at least 3. Is it true that*

$$\overline{mc}(G) \leq 2\chi(G) - 1?$$

Chapter 12
A Banquet Seating Problem

In this chapter, a banquet seating problem is described. The situation encountered in the problem can be naturally modeled by a graph and a vertex coloring of this graph. Such a vertex coloring, in turn, gives rise to other vertex colorings. The initial vertex coloring can be proper or not and, therefore, induces two different vertex colorings, which will be described in the following two chapters.

12.1 Seating Students at a Circular Table

A group of students (freshmen, sophomores, juniors, seniors, graduate students) have been invited to a banquet. Is it possible to seat all these students at a circular table in such a way that no two students belonging to the same class are seated next to two students belonging to the same class or the same two classes? For example, no two freshmen are to be seated next to two juniors, to two freshmen, to a senior and a graduate student, or to a freshmen and a sophomore. Consequently, every student is uniquely determined by the class to which he or she belongs and the classes of his or her two neighbors at the banquet table. Before considering examples of this type, let's look at some simpler examples.

At a banquet for students, a freshman, a sophomore and a junior are to be seated at a circular table. There is essentially only one way to do this since each student must sit next to students belonging to the other two classes. See Fig. 12.1.

Suppose next that two freshmen, one sophomore and one junior are to be seated at a circular table. There are two ways to do this. In the first seating arrangement of Fig. 12.2, each freshman sits next to the sophomore and junior. There is, therefore, no way to distinguish the two freshmen by the classes of the two students they sit next to. In the second seating arrangement of Fig. 12.2, one freshman sits next to the other freshman and the sophomore, while the other freshman sits next to a freshman and the junior. From this information, we see that these two freshmen can

© The Author 2016

P. Zhang, *A Kaleidoscopic View of Graph Colorings*, SpringerBriefs in Mathematics, DOI 10.1007/978-3-319-30518-9_12

Fig. 12.1 Seating a
freshman, a sophomore and a
junior

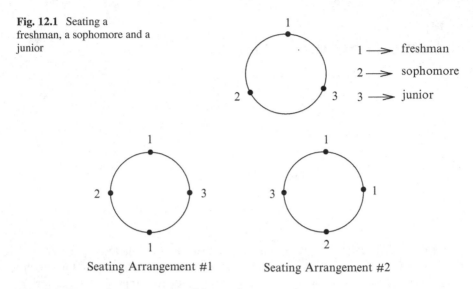

Seating Arrangement #1   Seating Arrangement #2

Fig. 12.2 Two seating arrangements of four students

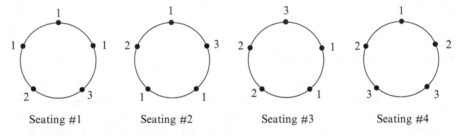

Seating #1   Seating #2   Seating #3   Seating #4

Fig. 12.3 Four seating arrangements of five students

be distinguished by the classes of the students they sit next to. In other words, once
we know the class of a student and the classes of the student's neighbors at the
table, we know exactly which student we are referring to. Thus, the second seating
arrangement satisfies the required conditions.

Suppose next that five students are to be seated at a circular table, each of whom
is a freshman, sophomore or junior. These students are to be seated so that the
neighbors of two students of the same class have different classes. The first three
seating arrangements in Fig. 12.3 satisfy these requirements; the fourth one does
not.

What if six or more students are to be seated at a circular table, each of whom
is a freshman, sophomore or junior? Figure 12.4 shows seating arrangements of six,
seven, eight and nine students. This gives rise to the following question.

*What is the maximum number of freshmen, sophomores and juniors who can be
seated at a circular table so that each student can be identified by his/her class and
that of the two neighbors sitting next to him/her?*

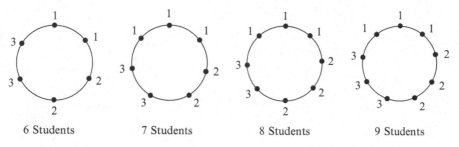

Fig. 12.4 Seating arrangements of six or more students

Fig. 12.5 A seating
arrangement of 18 students

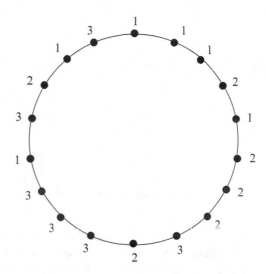

An example of seating 18 students, each of whom is a freshman, sophomore or junior at a circular table is shown in Fig. 12.5. Note that there are six freshmen, six sophomores and six juniors.

What if seniors as well are to seated at a circular table? Thus, we have the following question.

How many freshmen, sophomores, juniors and seniors can be seated at a circular table that would satisfy these conditions?

An example of seating 36 students, each of whom is a freshman, sophomore, junior or senior (where a senior is indicated by 4) at a circular table is shown in Fig. 12.6. There are nine students for each class of students.

What if graduate students are invited as well? We now have the following question.

How many freshmen, sophomores, juniors, seniors and graduate students can be seated at a circular table that would satisfy these conditions?

The largest possible number of freshmen, sophomores, juniors, seniors and graduate students who can be seated at a circular table with these conditions is 75.

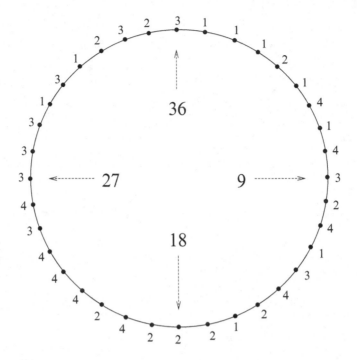

Fig. 12.6 A seating arrangement of 36 students

This is shown in Fig. 12.7, where a graduate student is indicated by 5. Therefore, if 75 students (15 freshmen, 15 sophomores, 15 juniors, 15 seniors, 15 graduate students) have been invited to a banquet, then it is possible to seat all 75 students at a 75-seat circular table in such a way that no 2 students belonging to the same class are seated next to 2 students belonging to the same class or the same two classes. Hence, every student is uniquely determined by the class to which he or she belongs and the classes of his or her two neighbors at the banquet table.

12.2   Modeling the Seating Problem by a Graph Coloring Problem

This banquet seating problem can be converted into a graph coloring problem. In this case, cycles are investigated and each vertex is assigned a color. For example, eight students of three different classes are seated at a circular table. Each student is represented by a vertex on the cycle and the class of the student becomes the color of the vertex. See Fig. 12.8.

More formally, for a graph G and a positive integer k, let

$$c : V(G) \to [k] = \{1, 2, \ldots, k\}$$

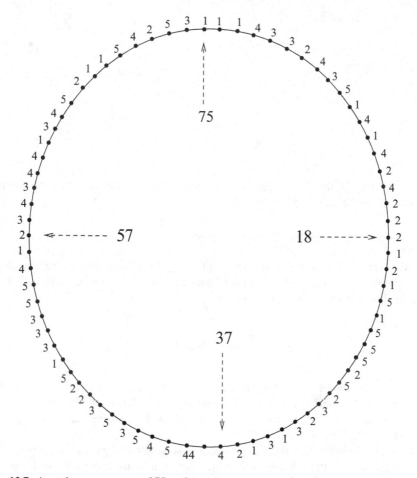

Fig. 12.7 A seating arrangement of 75 students

Fig. 12.8 A graph coloring

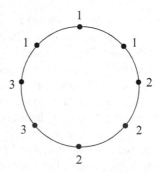

Fig. 12.9 Color codes of
vertices

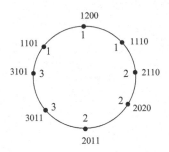

be a k-coloring of the vertices of G. Here the *color code* of a vertex v of G with
respect to c is the ordered $(k + 1)$-tuple

$$\text{code}(v) = (a_0, a_1, \ldots, a_k) = a_0 a_1 a_2 \cdots a_k,$$

where a_0 is the color assigned to v and a_i $(1 \le i \le k)$ is the number of vertices that
are adjacent to v and colored i. This is illustrated in Fig. 12.9 for the coloring of an
8-cycle shown in Fig. 12.8. Therefore,

$$\sum_{i=1}^{k} a_i = \deg_G v.$$

If distinct vertices of G have distinct color codes, then all vertices of G can be
distinguished by their color codes. For example, as shown in Fig. 12.9, no two
vertices of the 8-cycle have the same color code and so all vertices of this graph
are distinguished by their color codes.

Thus, the situation described in the banquet seating problem can be analyzed by
assigning colors to the vertices of a cycle such that distinct vertices of the cycle
have distinct color codes. In our examples, adjacent vertices can be assigned the
same colors.

This suggests studying colorings of the vertices of a graph where adjacent
vertices may be colored the same but where no two vertices have the same color
code. This also suggests another coloring problem, namely, that of studying *proper*
colorings of the vertices of a graph so that no two vertices assigned the same color
have the same number of neighbors assigned the same color for each color. We begin
by looking at 3-colorings of cycles.

This is equivalent to the problem of seating freshmen, sophomores and juniors at
a circular table such that (1) no two students of the same class sit next to each other
and (2) for every two students of the same class, their neighbors belong to different
classes. For example, Fig. 12.10 illustrates such a situation where an odd number of
students are seated at a circular table.

When four students are seated at a circular table and each of the students is a
freshman, sophomore or junior, there must be two students of the same class, say
two students are freshmen. They are not permitted to sit next to each other and so

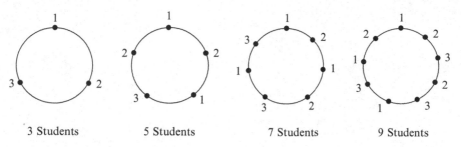

Fig. 12.10 Seating arrangements of 3, 5, 7, or 9 students

they must sit across the table from each other. The two vertices colored 1 have one neighbor colored 2 and one neighbor colored 3, and so both have color code 1011, which is not permitted.

As mentioned earlier, the problems described here give rise to two (different but related) graph coloring problems, both of which are investigated in the following two chapters.

Chapter 13
Irregular Colorings

In Chaps. 7–11, nonproper vertex colorings of graphs were described that gave rise to proper vertex colorings that may require fewer colors than the chromatic number of the graphs. In Sect. 12.1, a problem was described whose solution involved producing a vertex-distinguishing coloring of a cycle from a given vertex coloring (proper or nonproper). Here, we look at the situation where the given vertex coloring is proper, not just of cycles but graphs in general.

13.1 Irregular Chromatic Number

For a graph G and a positive integer k, let $c : V(G) \to [k] = \{1, 2, \ldots, k\}$ be a proper k-coloring of the vertices of G. Here, the *color code* (or simply the *code*) of a vertex v of G with respect to c is the ordered $(k + 1)$-tuple

$$\text{code}(v) = (a_0, a_1, \ldots, a_k) = a_0 a_1 a_2 \cdots a_k,$$

where $a_0 = c(v)$ and a_i $(1 \leq i \leq k)$ is the number of vertices that are adjacent to v and colored i. Consequently, if $c(v) = i$, then $a_i = 0$. Also,

$$\sum_{i=1}^{k} a_i = \deg_G v.$$

The coloring c is called *irregular* if distinct vertices of G have distinct color codes. The *irregular chromatic number* $\chi_{ir}(G)$ of G is the minimum positive integer k for which G has an irregular k-coloring. This concept was introduced by Mary Radcliffe

© The Author 2016
P. Zhang, *A Kaleidoscopic View of Graph Colorings*, SpringerBriefs in Mathematics,
DOI 10.1007/978-3-319-30518-9_13

and Ping Zhang [65] and studied further by others (see [2, 3, 64, 66], for example). Since every irregular coloring of a graph G is a proper coloring of G, it follows that

$$\chi(G) \leq \chi_{ir}(G). \tag{13.1}$$

The following useful observations were stated in [65].

Observation 13.1.1. *Let c be a proper vertex coloring of a nontrivial graph G and let u and v be two distinct vertices of G.*

(a) If $c(u) \neq c(v)$, then code$(u) \neq$ code(v).
(b) If $\deg_G u \neq \deg_G v$, then code$(u) \neq$ code(v).
(c) If c is irregular and $N(u) = N(v)$, then $c(u) \neq c(v)$.

To illustrate this concept, consider the Petersen graph P of Fig. 13.1. Since $\chi(P) = 3$, it follows by (13.1) that $\chi_{ir}(P) \geq 3$. A 4-coloring of the Petersen graph is given in Fig. 13.1 along with the corresponding color codes of its vertices. Since distinct vertices have distinct codes, this coloring is irregular and so $\chi_{ir}(P) \leq 4$. Therefore, $\chi_{ir}(P) = 3$ or $\chi_{ir}(P) = 4$. We show that $\chi_{ir}(P) = 4$. Assume, to the contrary, that $\chi_{ir}(P) = 3$. Then there exists an irregular 3-coloring c of P. Let u and v be two vertices of P with $c(u) = c(v)$. We may assume that $c(u) = c(v) = 1$. Since c is a proper coloring, u and v are not adjacent. Furthermore, the diameter of P is 2 and so u and v have a common neighbor in P. Because code$(u) \neq$ code(v), at most one of u and v is adjacent to three vertices having the same color. That is, no two vertices in P colored 1 can have the two color codes 1030 and 1003. Hence, if some vertex has color code 1030, then any other vertex colored 1 has color code 1021 or 1012. This implies that at most three vertices of P can be colored 1 and, in general, at most three vertices of P can be assigned the same color. Since P has order 10, this contradicts our assumption that $\chi_{ir}(P) = 3$.

Fig. 13.1 An irregular 4-coloring of the Petersen graph P

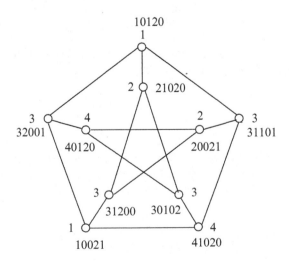

Recall that if A is a multiset containing ℓ different kinds of elements, where there are at least r elements of each kind, then the number of different selections of r elements from A is $\binom{r+\ell-1}{r}$ (see Sect. 1.5). We have seen that if c is an irregular k-coloring of a graph G and $v \in V(G)$ such that $\text{code}(v) = (i, a_1, \ldots, a_k)$ for some i with $1 \leq i \leq k$, then $a_i = 0$ and the sum of the remaining $k - 1$ coordinates a_1, $\ldots, a_{i-1}, a_{i+1}, \ldots, a_k$ is the degree of v. Therefore, by Theorem 1.5.1, we have the following.

Theorem 13.1.2 ([65]). *Let c be an irregular k-coloring of the vertices of a graph G. The number of different possible color codes of the vertices of degree r in G is*

$$k\binom{r + (k - 1) - 1}{r} = k\binom{r + k - 2}{r}.$$

The following result is a consequence of Theorem 13.1.2.

Corollary 13.1.3 ([65]). *If c is an irregular k-coloring of a nontrivial connected graph G, then G contains at most $k\binom{r+k-2}{r}$ vertices of degree r.*

In [3], Anderson, Vitray and Yellen showed that the result in Corollary 13.1.3 is best possible.

Because $\chi(G) \leq \chi_{ir}(G) \leq n$ for every graph G of order n and $\chi(K_n) = n$, it follows that $\chi_{ir}(K_n) = n$. The complete graph K_n is not the only graph of order n with irregular chromatic number n, however.

Theorem 13.1.4 ([65]). *Let G be a connected graph of order $n \geq 2$. Then $\chi_{ir}(G) = n$ if and only if $N(u) = N(v)$ for every pair u, v of nonadjacent vertices of G.*

With the aid of Theorem 13.1.4, all connected graphs G of order $n \geq 2$ with $\chi_{ir}(G) = n$ can be characterized.

Corollary 13.1.5 ([65]). *Let G be a connected graph of order $n \geq 2$. Then $\chi_{ir}(G) = n$ if and only if G is a complete multipartite graph.*

For a positive even integer $n = 2k$, let F_n be the bipartite graph with partite sets $X = \{x_1, x_2, \cdots, x_k\}$ and $Y = \{y_1, y_2, \ldots, y_k\}$ such that $\deg x_i = \deg y_i = i$ for $1 \leq i \leq k$. Figure 13.2 shows the graph F_8 of order 8 together with an irregular 2-coloring of F_8. It is easy to see that $\chi_{ir}(F_n) = 2$ for every even integer $n \geq 2$. In fact, more can be said.

Fig. 13.2 The graph F_8

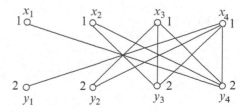

Theorem 13.1.6 ([65]). *Let G be a nontrivial connected graph of order n. Then*

$$\chi_{ir}(G) = 2 \text{ if and only if nis even and } G \cong F_n.$$

In [65] all pairs k, n of integers that can be realized as the irregular chromatic number and order of some connected graph are determined.

Theorem 13.1.7 ([65]). *For each pair k, n of integers with $2 \le k \le n$, there exists a connected graph of order n having irregular chromatic number k if and only if $(k, n) \ne (2, n)$ for any odd integer n.*

We have mentioned that $2 \le \chi(G) \le \chi_{ir}(G)$ for every nontrivial connected graph G. Other than these inequalities, there are no other restrictions on the values of the chromatic number and the irregular chromatic number of a nontrivial connected graph.

Corollary 13.1.8. *For every pair a, b of integers with $2 \le a \le b$, there is a connected graph G with*

$$\chi(G) = a \text{ and } \chi_{ir}(G) = b.$$

Proof. While the complete a-partite graph $G = K_{1,1,\ldots,1,b-a+1}$ has chromatic number a, it follows by Corollary 13.1.5 that $\chi_{ir}(G) = b$.   □

13.2   de Bruijn Sequences and Digraphs

In order to investigate the irregular chromatic numbers of cycles (which will be done in the following section), it is useful to become aware of a special sequence and a related digraph, both of which are described in this section. Prior to doing this, however, we make a few observations. Since $\chi(C_n) = 3$, where $n \ge 3$ is odd, $\chi_{ir}(C_n) \ge 3$ for all odd integers $n \ge 3$. The irregular 3-colorings of C_3, C_5 and C_7 in Fig. 13.3 show that the irregular chromatic number of all three cycles is 3.

On the other hand, even though $\chi(C_n) = 2$ for every even integer $n \ge 4$, no even cycle has irregular chromatic number 2 (or even 3). Since $C_4 = K_{2,2}$, it follows by Corollary 13.1.5 that $\chi_{ir}(C_4) = 4$. In fact, $\chi_{ir}(C_6) = \chi_{ir}(C_8) = 4$ as well. Irregular 4-colorings of these three cycles are shown in Fig. 13.4.

Fig. 13.3 Irregular
3-colorings of C_3, C_5 and C_7

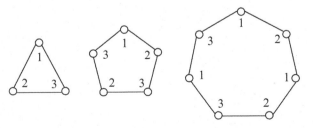

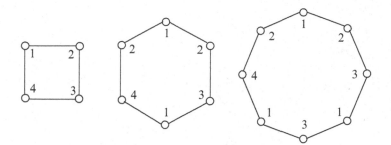

Fig. 13.4 Irregular 4-colorings of C_4, C_6 and C_8

Fig. 13.5 An irregular
3-coloring of C_9

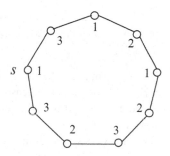

By Corollary 13.1.3, every nontrivial connected graph having an irregular 3-coloring contains at most nine vertices of degree 2. As Fig. 13.5 shows, there is an irregular 3-coloring of C_9. Consequently, $3 = \chi(C_9) \leq \chi_{ir}(C_9) \leq 3$ and so $\chi_{ir}(C_9) = 3$.

Although it is not all that challenging to construct an irregular 3-coloring of C_9, let's see how the particular irregular 3-coloring of C_9 shown in Fig. 13.5 can be constructed. To do this, we make use of a sequence and a digraph named for a Dutch mathematician. Nicolaas Govert de Bruijn was an innovative researcher who contributed to many areas of mathematics but he may be best known for a sequence and a digraph that bear his name.

Let A be a set consisting of $k \geq 2$ elements. For a positive integer n, an n-*word* over A is a sequence of length n whose terms belong to A. There are therefore k^n distinct n-words over A. A *de Bruijn sequence* is a sequence $a_0 a_1 \cdots a_{N-1}$ of elements of A having length $N = k^n$ such that for each n-word w over A, there is a unique integer i with $0 \leq i \leq N - 1$ such that $w = a_i a_{i+1} \cdots a_{i+n-1}$ where addition in the subscripts is performed modulo N.

For example, if $k = 3$ and $n = 2$ (so $A = \{0, 1, 2\}$), then $N = k^n = 3^2$ and the nine distinct 2-words over A are

$$00, \ 01, \ 11, \ 10, \ 02, \ 22, \ 21, \ 12, \ 20.$$

In fact, 001102212 is a de Bruijn sequence in this case.

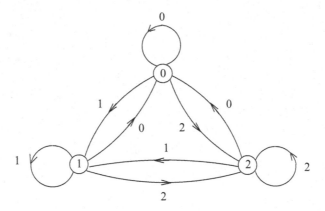

Fig. 13.6 The de Bruijn digraph $B(3, 2)$

While it may not be difficult to construct a de Bruijn sequence for small values of k and n, this is not the case when k or n is large. However, de Bruijn sequences can be constructed with the aid of a digraph (actually a digraph in which directed loops are permitted).

For integers $k, n \geq 2$, the *de Bruijn digraph* $B(k, n)$ is that digraph of order k^{n-1} whose vertex set is the set of $(n - 1)$-words over $A = \{0, 1, \cdots, k - 1\}$ and size k^n whose arc set consists of all n-words over A, where the arc $a_1 a_2 \cdots a_n$ is the ordered pair $(a_1 a_2 \cdots a_{n-1}, a_2 a_3 \cdots a_n)$ of vertices. Since the vertex $a_1 a_2 \cdots a_{n-1}$ is adjacent to the vertex $a_2 a_3 \cdots a_n$, we need only label the arc from $a_1 a_2 \cdots a_{n-1}$ to $a_2 a_3 \cdots a_n$ by a_n to indicate that the initial term a_1 is removed from $a_1 a_2 \cdots a_{n-1}$ and a_n is added as the final term to produce $a_2 a_3 \cdots a_n$. The de Bruijn digraph $B(3, 2)$ is shown in Fig. 13.6. Since the outdegree and indegree of every vertex of $B(3, 2)$ are equal, it follows that $B(3, 2)$ is Eulerian (see Sect. 1.4). One Eulerian circuit of $B(3, 2)$ is $(0, 0, 1, 1, 0, 2, 2, 1, 2, 0)$, which results in the de Bruijn sequence 001102212.

Because the de Bruijn digraph $B(k, n)$ is connected and the outdegree and indegree of every vertex of $B(k, n)$ is k, it follows that for every two integers $k, n \geq 2$, the de Bruijn digraph $B(k, n)$ is Eulerian (see Sect. 1.4). The de Bruijn digraph $B(2, 4)$ is shown in Fig. 13.7.

13.3   The Irregular Chromatic Numbers of Cycles

With the aid of de Bruijn digraphs, we now continue our investigation of the irregular chromatic numbers of cycles. We begin with an example. The vertex set of a subdigraph D of the de Bruijn digraph $B(3, 3)$ consists of the six 2-permutations of the set $\{1, 2, 3\}$, that is,

$$V(D) = \{12, 21, 13, 31, 23, 32\}.$$

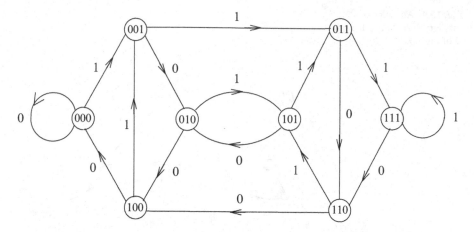

Fig. 13.7   The de Bruijn digraph $B(2,4)$

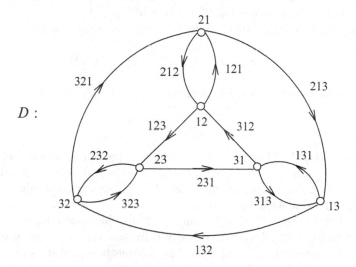

Fig. 13.8   A subdigraph of the de Bruijn digraph $B(3,3)$

Thus, $V(D)$ contains none of 11, 22 or 33. A vertex ab is *adjacent to* a vertex cd in D if $b = c$ and the resulting arc is labeled abd (or simply d). This subdigraph D of $B(3,3)$ is shown in Fig. 13.8.

The arc abd corresponds to a vertex of C_9 colored b and adjacent to vertices colored a and d in C_9. However, the vertex db in D is also adjacent to the vertex ba resulting in the arc dba and this also gives rise to a vertex in C_9 colored b adjacent to vertices colored d and a. In the desired irregular 3-coloring of C_9, only one vertex of C_9 can be colored b and adjacent to vertices colored a and d. Consequently, we seek a spanning Eulerian subdigraph D' of D containing only one of the arcs xyz or

Fig. 13.9 An Eulerian subdigraph D' of the digraph D of Fig. 13.8

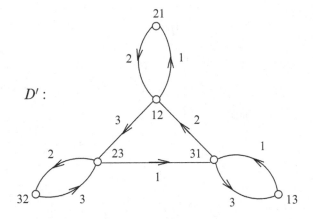

zyx if $x \neq z$. An Eulerian circuit C' of D' can then be used to produce an irregular 3-coloring of C_9. One such digraph D' is shown in Fig. 13.9.

Actually, the Eulerian circuit C' of the digraph D' is unique, namely

$$C' = (13, 31, 12, 21, 12, 23, 32, 23, 31, 13).$$

Following along the arcs of C' gives the irregular 3-coloring

$$1, 3, 1, 2, 1, 2, 3, 2, 3$$

of C_9 shown in Fig. 13.5 (starting at the vertex s and proceeding clockwise).

By Corollary 13.1.3, $\chi_{ir}(C_n) \geq 4$ for every integer $n \geq 10$ and the largest integer n for which $\chi_{ir}(C_n)$ can have the value 4 is 24. The irregular chromatic number of C_{24} is in fact 4, as shown in Fig. 13.10.

The fact that $\chi_{ir}(C_{24}) = 4$ has an interesting interpretation. Suppose, for example, that there are 24 students (namely 6 freshmen, 6 sophomores, 6 juniors and 6 seniors) attending a banquet. Can all 24 students be seated at a single circular table in such a way that no 2 students from the same class are seated next to each other and no 2 students from the same class have neighbors from the same class or pair of classes? Since $\chi_{ir}(C_{24}) = 4$, this question has an affirmative answer. In fact, the irregular 4-coloring of C_{24} in Fig. 13.10 gives a possible seating arrangement (where 1 represents a freshman, 2 a sophomore, 3 a junior and 4 a senior).

A formula for the irregular chromatic number of all cycles was obtained by Anderson et al. [2].

Theorem 13.3.1. *Let $k \geq 4$. If $(k-1)\binom{k-1}{2} + 1 \leq n \leq k\binom{k}{2}$, then*

$$\chi_{ir}(C_n) = \begin{cases} k & \text{if } n \neq k\binom{k}{2} - 1 \\ k+1 & \text{if } n = k\binom{k}{2} - 1. \end{cases}$$

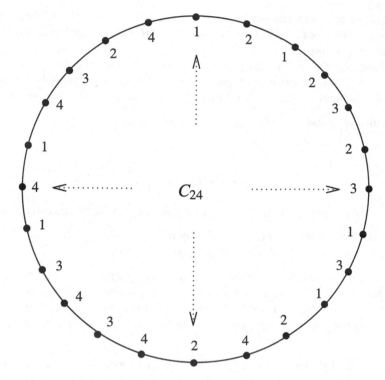

Fig. 13.10 An irregular 4-coloring of C_{24}

13.4 Nordhaus-Gaddum Inequalities

We close this chapter by presenting a result on irregular chromatic numbers that is an analogue of a well-known theorem on chromatic numbers. Recall that the *complement* $\overline{G}$ of a graph G is that graph with vertex set $V(G)$ such that two vertices are adjacent in $\overline{G}$ if and only if these vertices are not adjacent in G. Probably the best known result involving a graph and its complement provides upper and lower bounds for both the sum and product of the chromatic numbers of a graph and its complement. The following theorem is due to Edward A. Nordhaus and Jerry W. Gaddum.

Theorem 13.4.1 ([52]). *For every graph G of order n,*

$$2\sqrt{n} \leq \chi(G) + \chi(\overline{G}) \leq n + 1$$

$$n \leq \chi(G)\chi(\overline{G}) \leq \left(\frac{n+1}{2}\right)^2.$$

Indeed for any graphical parameter f and any graph G of order n, sharp upper and lower bounds for both $f(G) + f(\overline{G})$ and $f(G) \cdot f(\overline{G})$ are referred to as *Nordhaus-Gaddum inequalities*.

We have seen that $2 \leq \chi_{ir}(G) \leq n$ and $\chi_{ir}(G) \geq \chi(G)$ for every nontrivial graph G of order n. These observations together with Theorem 13.4.1 yield the following Nordhaus-Gaddum inequalities for the irregular chromatic number of a graph.

Theorem 13.4.2 ([64]). *If G is a graph of order n, then*

$$2\sqrt{n} \leq \chi_{ir}(G) + \chi_{ir}(\overline{G}) \leq 2n$$

$$n \leq \chi_{ir}(G)\chi_{ir}(\overline{G}) \leq n^2.$$

Each of the four bounds in Theorem 13.4.2 is sharp. In fact, more can be said.

Theorem 13.4.3 ([64]). *Let G be a graph of order n. Then $\chi_{ir}(G) + \chi_{ir}(\overline{G}) = 2n$ or $\chi_{ir}(G)\chi_{ir}(\overline{G}) = n^2$ if and only if $G = K_n$ or $G = \overline{K}_n$.*

To see that the lower bound in the inequality $\chi_{ir}(G)\chi_{ir}(\overline{G}) \geq n$ in Theorem 13.4.2 is sharp, recall that for each even integer $n = 2k$, where k is a positive integer, the bipartite graph F_n described in Theorem 13.1.6 is the only connected graph of order n with irregular chromatic number 2. In fact, $\chi_{ir}(\overline{F}_n) = k$ and so we have the following.

Theorem 13.4.4 ([64]). *For every positive even integer n, there exists a connected graph G of order n such that $\chi_{ir}(G)\chi_{ir}(\overline{G}) = n$.*

The lower bound in the inequality $\chi_{ir}(G) + \chi_{ir}(\overline{G}) \geq 2\sqrt{n}$ in Theorem 13.4.2 is also sharp. Observe that if G is a nontrivial graph of order n such that $\chi_{ir}(G) + \chi_{ir}(\overline{G}) = 2\sqrt{n}$, then obviously $n = p^2$ for some positive integer p.

Theorem 13.4.5 ([64]). *For every positive integer p, there exists a graph G of order $n = p^2$ such that $\chi_{ir}(G) + \chi_{ir}(\overline{G}) = 2\sqrt{n}$.*

A point (a, b) in the plane is called a *lattice point* if a and b are integers. If G is a graph of order n such that $\chi_{ir}(G) = a$ and $\chi_{ir}(\overline{G}) = b$, then $2\sqrt{n} \leq a + b \leq 2n$ and $n \leq ab \leq n^2$, which is equivalent to $n \leq ab$ and $a + b \leq 2n$. For this reason, we define a lattice point (a, b) to be *realizable* with respect to an integer n if $n \leq ab$ and $a + b \leq 2n$ and there is a graph G of order n such that $\chi_{ir}(G) = a$ and $\chi_{ir}(\overline{G}) = b$. For the chromatic number, Hans-Joachim Finck [36] and Bonnie M. Stewart [70] showed that no improvement in Theorem 13.4.1 is possible (without employing additional conditions). We state this theorem as follows.

Theorem 13.4.6 ([36, 70]). *Let n be a positive integer. For every two integers a and b such that*

$$2\sqrt{n} \leq a + b \leq n + 1, \text{ and } n \leq ab \leq \left(\frac{n+1}{2}\right)^2$$

there is a graph G of order n such that $\chi(G) = a$ and $\chi(\overline{G}) = b$.

There is, however, no corresponding result for irregular chromatic numbers. We conclude this section with the following question.

Question 13.4.7. *Let $n \geq 2$ be an integer. For which lattice points (a, b) with*

$$a, b \geq 2, \ n \leq ab \ and \ a + b \leq 2n,$$

does there exist a graph G of order n such that

$$\chi_{ir}(G) = a \ and \chi_{ir}(\overline{G}) = b?$$

Chapter 14
Recognizable Colorings

In the preceding chapter, irregular colorings of graphs were considered. These are proper vertex colorings $c : V(G) \to [k] = \{1, 2, \ldots, k\}$ that give rise to a vertex-distinguishing coloring of G whose colors are $(k+1)$-tuples of nonnegative integers. In this chapter, we discuss the corresponding $(k + 1)$-tuples when the original coloring is a nonproper coloring. This gives rise to vertex-distinguishing colorings called recognizable colorings.

14.1   The Recognition Numbers of Graphs

Let G be a graph and let $c : V(G) \to [k] = \{1, 2, \ldots, k\}$ be a coloring of the vertices of G for some positive integer k (where adjacent vertices may be colored the same). The *color code* of a vertex v of G (with respect to c) is the ordered $(k + 1)$-tuple

$$\text{code}_c(v) = (a_0, a_1, \ldots, a_k) \ (\text{or simply}, \text{code}(v) = a_0 a_1 a_2 \cdots a_k),$$

where a_0 is the color assigned to v (that is, $c(v) = a_0$) and for $1 \leq i \leq k$, a_i is the number of vertices adjacent to v that are colored i. Therefore, $\sum_{i=1}^{k} a_i = \deg_G v$. The coloring c is called *recognizable* if distinct vertices have distinct color codes and the *recognition number* $\text{rn}(G)$ of G is the minimum positive integer k for which G has a recognizable k-coloring. A recognizable coloring of G using $\text{rn}(G)$ colors is called a *minimum recognizable coloring*. These concepts were introduced in [19]. A graph G and its complement $\overline{G}$ have the same recognition number.

Proposition 14.1.1. *For every graph G, $\text{rn}(G) = \text{rn}(\overline{G})$.*

Proof. Suppose that $\text{rn}(G) = k$ and $\text{rn}(\overline{G}) = \overline{k}$. Let c be a recognizable k-coloring of G. Define a k-coloring $\overline{c}$ of $\overline{G}$ by $\overline{c}(v) = c(v)$ for each $v \in V(\overline{G}) = V(G)$. Suppose, in the coloring c of G, there are n_i vertices of G colored i for $1 \leq i \leq k$.

© The Author 2016

P. Zhang, *A Kaleidoscopic View of Graph Colorings*, SpringerBriefs in Mathematics,
DOI 10.1007/978-3-319-30518-9_14

Let x and y be two vertices of $\overline{G}$ that have same color code with respect to $\overline{c}$. We may assume that $\overline{c}(x) = \overline{c}(y) = 1$ and that

$$\mathrm{code}_{\overline{c}}(x) = (1, a_1, a_2 \ldots, a_k) = \mathrm{code}_{\overline{c}}(y).$$

Consequently,

$$\mathrm{code}_c(x) = (1, n_1 - a_1 - 1, n_2 - a_2, \ldots, n_k - a_k) = \mathrm{code}_c(y).$$

Since c is recognizable, $x = y$, which implies that $\overline{c}$ is a recognizable k-coloring of $\overline{G}$. Since $\mathrm{rn}(\overline{G}) = \overline{k}$, it follows that $\overline{k} \leq k$. By a similar argument, $k \leq \overline{k}$ and so $k = \overline{k}$. Thus, $\mathrm{rn}(G) = \mathrm{rn}(\overline{G})$.   □

Since the complement of every disconnected graph is connected, it follows by Proposition 14.1.1 that the study of recognizable colorings of graphs can be restricted to connected graphs. Every coloring that assigns distinct colors to the vertices of a connected graph is recognizable; so the recognition number is always defined. On the other hand, it is well-known that every nontrivial graph contains at least two vertices having the same degree. Thus, if all vertices of a nontrivial graph are assigned the same color, then any two vertices of the same degree will have the same color code. Therefore, if G is a nontrivial connected graph of order n, then

$$2 \leq \mathrm{rn}(G) \leq n. \tag{14.1}$$

There are two observations that will be useful to us.

Observation 14.1.2. *Let c be a coloring of the vertices of a graph G. If u and v are two vertices of G with $\deg_G u \neq \deg_G v$, then $\mathrm{code}(u) \neq \mathrm{code}(v)$.*

In particular, to show that a coloring of a graph G is recognizable, it is necessary and sufficient to show that every two vertices of the same degree and same color have distinct codes.

Observation 14.1.3. *Let c be a recognizable coloring of a graph G. If u and v are distinct vertices of G with $N[u] = N[v]$, then $c(u) \neq c(v)$.*

By Observation 14.1.3, the complete graph K_n of order n has recognition number n. In fact, it is the only connected graph of order n having this property.

Proposition 14.1.4 ([19]). *If G is a nontrivial connected graph of order n, then*

$$\mathrm{rn}(G) = n \text{ if and only if } G = K_n.$$

Proof. Since $\mathrm{rn}(K_n) = n$ for each integer $n \geq 2$, it remains to show that K_n is the only connected graph of order n with recognition number n. Suppose that G is a connected graph of order n that is not complete. Thus, G contains three vertices x, y and z such that $xy \notin E(G)$ and $yz \in E(G)$. Define a coloring c that assigns color 1 to

Fig. 14.1 A minimum recognizable coloring of the Petersen graph

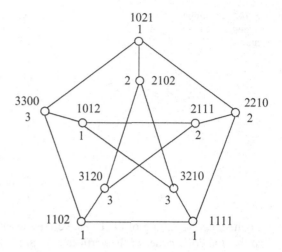

x and z, color 2 to y and distinct colors from the set $\{3, 4, \ldots, n-1\}$ to the remaining $n - 3$ vertices of G. Since c is a recognizable $(n - 1)$-coloring of G, it follows that $rn(G) \leq n - 1$. □

Recall, once again, that if A is a multiset containing k different kinds of elements, where there are at least r elements of each kind, then the number of different selections of r elements from A is $\binom{r+k-1}{r}$ (see Sect. 1.5). The following is a consequence of this result.

Theorem 14.1.5. *If c is a recognizable k-coloring of a nontrivial connected graph G, then G contains at most $k\binom{r+k-1}{r}$ vertices of degree r.*

We now consider examples of recognizable colorings of some cubic graphs. The following is a consequence of Theorem 14.1.5 for cubic graphs.

Corollary 14.1.6. *If G is a connected cubic graph of order n having recognition number k, then*

$$n \leq \frac{k^4 + 3k^3 + 2k^2}{6}.$$

The Petersen graph P is a cubic graph of order 10. By Corollary 14.1.6, $rn(P) \geq 3$. A 3-coloring of the Petersen graph is given in Fig. 14.1 along with the corresponding color codes of its vertices. Since distinct vertices have distinct color codes, this coloring is recognizable. Thus, $rn(P) \leq 3$ and so $rn(P) = 3$.

According to Corollary 14.1.6, if G is a connected cubic graph of order n having recognition number 2, then $n \leq 8$. In fact, there is no connected cubic graph of order 8 with recognition number 2, however. To see this, assume, to the contrary, that there exists a cubic graph G of order 8 with $rn(G) = 2$. Therefore, there is a

Fig. 14.2 A minimum
recognizable coloring of the
3-cube

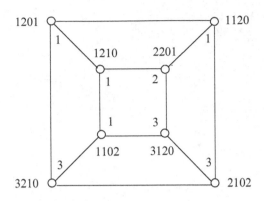

recognizable 2-coloring of G. Thus, we may assume that $V(G) = \{v_1, v_2, \ldots, v_8\}$
and that the color codes of the vertices of G are

$$\text{code}(v_1) = 130, \text{code}(v_2) = 103, \text{code}(v_3) = 121, \text{code}(v_4) = 112,$$

$$\text{code}(v_5) = 230, \text{code}(v_6) = 203, \text{code}(v_7) = 221, \text{code}(v_8) = 212.$$

Since v_1 is colored 1 and is adjacent to three vertices colored 1, namely v_2, v_3 and
v_4, it follows that each of v_2, v_3 and v_4 is adjacent to at least one vertex colored 1.
Since $\text{code}(v_2) = 103$, the vertex v_2 is adjacent to no vertex colored 1, producing a
contradiction.

A well-known cubic graph of order 8 is the 3-cube Q_3, which is shown in
Fig. 14.2. Since there is no connected cubic graph of order 8 with recognition
number 2, it follows that $\text{rn}(Q_3) \geq 3$. A 3-coloring of Q_3 is given in Fig. 14.2
along with the corresponding color codes of its vertices. Since distinct vertices
have distinct color codes, this coloring is recognizable. Thus, $\text{rn}(Q_3) \leq 3$ and so
$\text{rn}(Q_3) = 3$. Hence, there is a cubic graph with recognition number 3.

14.2   Complete Multipartite Graphs

In this section we determine the recognition numbers of all complete multipartite
graphs. Let G be a complete k-partite graph for some positive integer k. Recall that
if every partite set of G has a vertices for some positive integer a, then we write
$G = K_{k(a)}$, where then $K_{1(a)} = \overline{K}_a$.

Theorem 14.2.1 ([19]). *Let k and a be positive integers. Then the recognition
number of the complete k-partite graph $K_{k(a)}$ is the unique positive integer ℓ for
which*

$$\binom{\ell - 1}{a} < k \leq \binom{\ell}{a}.$$

Proof. Suppose that $G = K_{k(a)}$ has partite sets $U_1, U_2, \ldots, U_k$, where $|U_i| = a$ for $1 \leq i \leq k$. We first show that $\mathrm{rn}(G) \geq \ell$. Assume, to the contrary, that $\mathrm{rn}(G) \leq \ell - 1$. Then there exists a recognizable coloring c of G using $\ell - 1$ or fewer colors. Let $\mathcal{S} = [\ell - 1]$. For each integer i with $1 \leq i \leq k$, let

$$\mathcal{C}_i = \{c(x) : x \in U_i\}$$

be the set of the colors of the vertices of U_i. Then $\mathcal{C}_i$ is an a-element subset of S for $1 \leq i \leq k$. Since S has exactly $\binom{\ell-1}{a}$ distinct a-element subsets and $k > \binom{\ell-1}{a}$, it follows that there exist two partite sets U_s and U_t, where $1 \leq s \neq t \leq k$, such that $\mathcal{C}_s = \mathcal{C}_t$, that is,

$$\{c(x) : x \in U_s\} = \{c(y) : y \in U_t\}.$$

Thus, there exist $x \in U_s$ and $y \in U_t$ such that $c(x) = c(y)$. However then, $\mathrm{code}(x) = \mathrm{code}(y)$, which contracts the fact that c is a recognizable coloring of G. Therefore, $\mathrm{rn}(G) \geq \ell$.

Next, we show that $\mathrm{rn}(G) \leq \ell$. Let $\mathcal{L} = [\ell]$ and let

$$\mathcal{L}_1, \mathcal{L}_2, \ldots, \mathcal{L}_{\binom{\ell}{a}}$$

be the $\binom{\ell}{a}$ distinct a-element subsets of L. Since $k \leq \binom{\ell}{a}$, we can define a coloring c' of G that assigns the a distinct colors of $\mathcal{L}_i$ to the vertices of U_i for $1 \leq i \leq k$. Since $k > \binom{\ell-1}{a}$, there must be at least one vertex of G that is assigned color j for each color j with $1 \leq j \leq \ell$. Thus, c' is an ℓ-coloring of G. It remains to show that c' is a recognizable coloring. Let u and v be two vertices of G such that $c'(u) = c'(v)$. Since the a vertices in each partite set U_i are colored differently by c' for $1 \leq i \leq k$, it follows that u and v belong two different partite sets of G. We may assume, without loss of generality, that $u \in U_1$ and $v \in U_2$. Suppose that

$$\mathcal{C}'_1 = \{c'(x) : x \in U_1\} = \{s_1, s_2, \ldots, s_a\}$$

$$\mathcal{C}'_2 = \{c'(y) : y \in U_2\} = \{t_1, t_2, \ldots, t_a\}.$$

Since $\mathcal{C}'_1 \neq \mathcal{C}'_2$, either there is an element in $\mathcal{C}'_1$ that is not in $\mathcal{C}'_2$ or an element in $\mathcal{C}'_2$ that is not in $\mathcal{C}'_1$, say the former. We may assume, without loss of generality, that $s_1 \notin \mathcal{C}'_2$. Let w be the vertex colored s_1 in U_1. (Note that it is possible that $w = u$.) Observe that

(1) u and v are both adjacent to every vertex in $V(G) - (U_1 \cup U_2)$ and so u and v are both adjacent to every vertex colored s_1 in $V(G) - (U_1 \cup U_2)$;
(2) v is adjacent to every vertex in U_1 but u is adjacent to no vertex in U_1, and so v is adjacent to the only vertex colored s_1 in $U_1 \cup U_2$, namely the vertex w, while u is not adjacent to w.

Thus, v is adjacent to every vertex colored s_1 in G, while u is adjacent to every vertex colored s_1 in G except w. Therefore, v is adjacent to exactly one more vertex colored s_1 in G than u is, and so the $(s_1 + 1)$st coordinate in code(v) does not equal the $(s_1 + 1)$st coordinate in code(u). Thus, code$(u) \neq$ code(v). Hence, c' is a recognizable ℓ-coloring of G and so rn$(G) \leq \ell$. Therefore, rn$(G) = \ell$.   □

In particular, if $a = 2$, then by solving $\ell^2 - \ell - 2t = 0$ for ℓ, we obtain $\ell = \frac{1+\sqrt{1+8t}}{2}$ and so

$$ \text{rn}(K_{t(2)}) = \left\lceil \frac{1 + \sqrt{1 + 8t}}{2} \right\rceil. $$

Recall that if a complete multipartite graph G contains t_i partite sets of cardinality n_i for every integer i with $1 \leq i \leq k$, then we write $G = K_{t_1(n_1), t_2(n_2), \cdots, t_k(n_k)}$.

Corollary 14.2.2 ([19]). *Let* $G = K_{t_1(n_1), t_2(n_2), \cdots, t_k(n_k)}$, *where* $n_1, n_2, \ldots, n_k$ *are* k *distinct positive integers. Then* rn$(G) = \max\{\text{rn}(K_{t_i(n_i)}) : 1 \leq i \leq k\}$.

Proof. Let $\ell_i = \text{rn}(K_{t_i(n_i)})$ for $1 \leq i \leq k$. Assume, without loss of generality, that

$$ \ell_1 = \max\{\text{rn}(K_{t_i(n_i)}) : 1 \leq i \leq k\}. $$

We first show that rn$(G) \leq \ell_1$. For each integer i with $1 \leq i \leq k$, let c_i be a recognizable ℓ_i-coloring of the subgraph $K_{t_i(n_i)}$ in G. We can now define a recognizable ℓ_1-coloring c of G by defining

$$ c(x) = c_i(x) \text{ if } x \in V(K_{t_i(n_i)}) \text{ for } 1 \leq i \leq k. $$

Thus, rn$(G) \leq \ell_1$. Next, we show that rn$(G) \geq \ell_1$. Assume, to the contrary, that rn$(G) = \ell \leq \ell_1 - 1$. Let c' be a recognizable ℓ-coloring of G. Then c' induces a coloring c_1' of the subgraph $K_{t_1(n_1)}$ in G such that $c_1'(x) = c(x)$ for all $x \in V(K_{t_1(n_1)})$. Since c_1' uses at most ℓ colors and rn$(K_{t_1(n_1)}) = \ell_1 > \ell$, it follows that c_1' is not a recognizable coloring of $K_{t_1(n_1)}$, and so there exist two vertices u and v in $K_{t_1(n_1)}$ such that u and v have the same color code with respect to c_1'. Since u and v are both adjacent to every vertex in $V(G) - V(K_{t_1(n_1)})$, it follows that u and v have the same color code in G with respect to c', which is a contradiction.   □

In particular, if $t_1 = t_2 = \cdots = t_k = 1$, then $K_{t_i(n_i)} = K_{1(n_i)} = \overline{K}_{n_i}$ for $1 \leq i \leq k$. Since rn$(\overline{K}_{n_i}) = n_i$ for $1 \leq i \leq k$, it follows that

$$ \text{rn}(K_{n_1, n_2, \ldots, n_k}) = \max\{n_i : 1 \leq i \leq k\}, $$

where $n_1, n_2, \ldots, n_k$ are k distinct positive integers. Furthermore, for integers s and t with $1 \leq s \leq t$,

$$ \text{rn}(K_{s,t}) = \begin{cases} t & \text{if } s < t, \\ t+1 & \text{if } s = t. \end{cases} $$

14.3   Graphs with Prescribed Order and Recognition Number

We saw in Proposition 14.1.4 that the complete graph K_n of order n is the only connected graph of order n having recognition number n. In this section we first state a characterization those connected graphs of order n having recognition number n or $n - 1$.

Theorem 14.3.1 ([19]). *Let G be a connected graph of order $n \geq 4$. Then*

$$\mathrm{rn}(G) = n - 1 \text{ if and only if } G = K_{1,n-1} \text{ or } G = C_4.$$

We have seen in (14.1) that if G is a nontrivial connected graph of order n having recognition number k, then $2 \leq k \leq n$. Next we show that every pair k, n of integers with $2 \leq k \leq n$ is realizable as the recognition number and order of some connected graph. The following observation will be useful in the proof of the next result.

Observation 14.3.2. *If G is a nontrivial connected graph such that the maximum number of vertices of the same degree is k, then $\mathrm{rn}(G) \leq k$.*

Theorem 14.3.3 ([19]). *For each pair k, n of integers with $2 \leq k \leq n$, there exists a connected graph of order n having recognition number k.*

Proof. For $k = 2$, let G be the unique connected graph of order n containing exactly two vertices of equal degree. It then follows by Observation 14.3.2 that $\mathrm{rn}(G) = 2$. For $k = n$, let $G = K_n$ and $\mathrm{rn}(K_n) = n$. If $k > n - k$, then let $G = K_{n-k,k}$ and $\mathrm{rn}(K_{n-k,k}) = k$. Thus, we may assume that $3 \leq k \leq n - k$. We consider two cases, according to whether $k \geq 4$ or $k = 3$.

Case 1. $k \geq 4$. Then $3 < k \leq n - k \leq n - 4$ and so $n - k \geq 4$. Let F be the unique connected graph of order $n - k$ containing exactly two vertices of equal degree. Then the degrees of the vertices of F are $1, 2, \cdots, \lfloor \frac{n-k}{2} \rfloor, \lfloor \frac{n-k}{2} \rfloor, \ldots, n - k - 1$. Let $V(F) = \{u_1, u_2, \ldots, u_{n-k}\}$, where $\deg_F u_1 = 1$ and $\deg_F u_{n-k} = n - k - 1$. Since $n - k \geq 4$, it follows that $2 \leq \lfloor \frac{n-k}{2} \rfloor < n - k - 1$ and so F has a unique end-vertex, namely u_1. The graph G is now constructed from F by adding k new vertices $v_1, v_2, \ldots, v_k$ and joining each vertex v_i ($1 \leq i \leq k$) to u_{n-k}. Then the order of G is n. It remains to show that $\mathrm{rn}(G) = k$. Since the k vertices $v_1, v_2, \ldots, v_k$ have the same closed neighborhood, it follows by Observation 14.1.3 that $\mathrm{rn}(G) \geq k$. On the other hand, the maximum number of vertices of the same degree is k. It then follows by Observation 14.3.2 that $\mathrm{rn}(G) \leq k$. Therefore, $\deg(G) = k$.

Case 2. $k = 3$. Since $3 = k \leq n - k = n - 3$, it follows that $n \geq 6$. Let F be the unique connected graph of order $n - 2$ containing exactly two vertices of equal degree. Then the degrees of the vertices of F are $1, 2, \cdots, \lfloor \frac{n-2}{2} \rfloor, \lfloor \frac{n-2}{2} \rfloor, \ldots, n - 3$. Let $V(F) = \{u_1, u_2, \ldots, u_{n-2}\}$, where $\deg_F u_1 = 1$ and $\deg_F u_{n-2} = n - 3$. Since $n - 2 \geq 4$, it follows that $\lfloor \frac{n-2}{2} \rfloor \geq 2$ and so u_1 is the unique end-vertex in F. Let v be the vertex adjacent to u_1 in F.

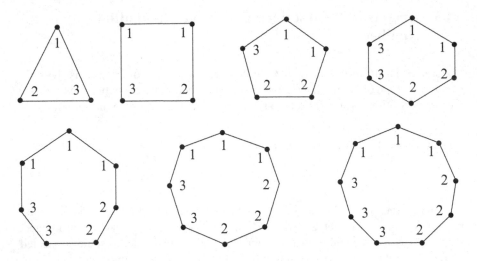

Fig. 14.3 Recognizable 3-colorings of cycles C_n, $3 \le n \le 9$

Now the graph G is obtained from F by adding two new vertices v_1 and v_2 and joining each of v_1 and v_2 to v. Then the order of G is n. It remains to show that $\mathrm{rn}(G) = 3$. Since the three end-vertices u_1, v_1, v_2 have the same neighborhood, $\mathrm{rn}(G) \ge 3$. On the other hand, the maximum number of vertices of the same degree is 3. It then follows by Observation 14.3.2 that $\mathrm{rn}(G) \le 3$. Therefore, $\deg(G) = 3$.   □

14.4   Recognizable Colorings of Cycles and Paths

It is straightforward to show that no 2-coloring of any cycle is recognizable. Therefore, $\mathrm{rn}(C_n) \ge 3$ for every integer $n \ge 3$. There are many cycles having recognition number 3, however. Recognizable 3-colorings of the cycles C_n, $3 \le n \le 9$, are shown in Fig. 14.3.

By Theorem 14.1.5, the number of vertices of degree 2 in a graph having recognition number k is at most $(k^3 + k^2)/2$. In particular, if $\mathrm{rn}(C_n) = 3$, then $n \le 18$. Since the 3-coloring of C_{18} shown in Fig. 14.4 (also in Fig. 12.5) is recognizable, it follows that $\mathrm{rn}(C_{18}) = 3$.

Again, by Theorem 14.1.5, if $\mathrm{rn}(C_n) = 4$, then $n \le 40$. There is no recognizable 4-coloring of C_{40}, however, for assume, to the contrary, that such a 4-coloring c of C_{40} exists. Then each of the 40 possible 5-tuples in this case is the color code of exactly one vertex of C_{40}. We show, however, that it is impossible for three vertices of C_{40} to have the color codes 11100, 11010, 11001; for suppose that there are three vertices of C_{40} with these color codes. Let $C_{40} = (v_1, v_2, \ldots, v_{40}, v_1)$, where $\mathrm{code}(v_2)$, say, is one of 11100, 11010 or 11001. We may assume that

Fig. 14.4 A recognizable
3-coloring of C_{18}

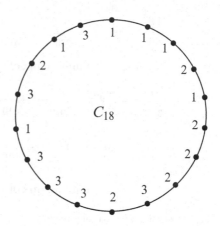

code(v_2) $= 11100$ and that $c(v_1) = 2$ and $c(v_2) = c(v_3) = 1$. Hence, code(v_3)
is either 11010, 11001 or 12000. If code(v_3) $= 12000$, then code(v_4) is either
11010 or 11001. Thus, if code(v_2) $= 11100$, then either code(v_3) or code(v_4) is
either 11010 or 11001, say 11010. However then, there can be no vertex v_i with
code(v_i) $= 11001$, for otherwise, either a neighbor of v_i or a vertex at distance 2
from v_i on C_{40} would have 11100 or 11010 as its color code, which is impossible.
Hence, rn(C_{40}) $\neq 4$. By the same argument, it is also impossible for three vertices of
C_n ($37 \leq n \leq 40$) to have (1) the color codes 21100, 20110 or 20101, (2) the color
codes 31010, 30110 or 30011 or (3) the color codes 41001, 40101 or 40011. Hence,
if rn(C_n) $= 4$, then $n \leq 36$. Therefore, at most 36 of the 40 possible 5-tuples in this
case can be color codes. Thus, if rn(C_n) $= 4$, then $n \leq 36$. In fact, rn(C_{36}) $= 4$, as
there is a recognizable 4-coloring of C_{36} (see Fig. 12.6). In general, there is a lower
bound for rn(C_n).

Theorem 14.4.1 ([19]). *Let $k \geq 3$ be an integer. Then* rn(C_n) $\geq k$ *for all integers n
such that*

$$\frac{(k-1)^3+(k-1)^2-2(k-1)+2}{2} \leq n \leq \frac{k^3+k^2}{2} \qquad \text{if } k \text{ is odd}$$

$$\frac{(k-1)^3+(k-1)^2+2}{2} \leq n \leq \frac{k^3+k^2-2k}{2} \qquad \text{if } k \text{ is even.}$$

As a result of Theorem 14.4.1 and the fact that it is possible to have two additional
color codes for vertices of degree 2 in the path P_n of order n than for C_n when k is
even, we have the following lower bound for rn(P_n).

Theorem 14.4.2 ([19]). *Let $k \geq 3$ be an integer. Then* rn(P_n) $\geq k$ *for all integers n
such that*

$$\frac{(k-1)^3+(k-1)^2-2(k-1)+10}{2} \leq n \leq \frac{k^3+k^2+4}{2} \qquad \text{if } k \text{ is odd}$$

$$\frac{(k-1)^3+(k-1)^2+6}{2} \leq n \leq \frac{k^3+k^2-2k+8}{2} \qquad \text{if } k \text{ is even.}$$

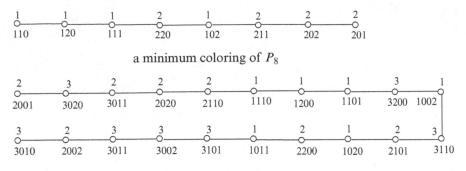

a minimum coloring of P_8

a minimum coloring of P_{20}

Fig. 14.5 Minimum colorings for P_8 and P_{20}

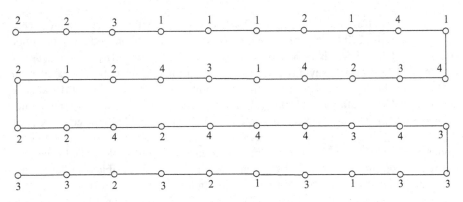

Fig. 14.6 A minimum recognizable 4-coloring of P_{40}

It is known that $rn(P_n) = 2$ if $2 \leq n \leq 8$ and $rn(P_n) = 3$ if $9 \leq n \leq 20$. For example, minimum recognizable colorings for P_8 and P_{20} are shown in Fig. 14.5 along with the corresponding color codes of their vertices.

The largest integer n for which $rn(P_n) = 4$ is $n = 40$. A recognizable 4-coloring of P_{40} is shown in Fig. 14.6 and so $rn(P_{40}) = 4$. The largest possible integer n for which $rn(P_n) = 5$ is $n = 77$. It can be shown that $rn(P_{77}) = 5$.

It was conjectured that the lower bound for $rn(C_n)$ in Theorem 14.4.1 and the lower bound for $rn(P_n)$ in Theorem 14.4.2 are equalities throughout, as we state next.

Conjecture 14.4.3 ([19]). *Let $k \geq 3$ be an integer. Then* $rn(C_n) = k$ *for all integers* n *such that*

$$\frac{(k-1)^3+(k-1)^2-2(k-1)+2}{2} \leq n \leq \frac{k^3+k^2}{2} \quad \text{if } k \text{ is odd}$$

$$\frac{(k-1)^3+(k-1)^2+2}{2} \leq n \leq \frac{k^3+k^2-2k}{2} \quad \text{if } k \text{ is even}$$

Conjecture 14.4.4 ([19]). *Let $k \geq 3$ be an integer. Then* $\mathrm{rn}(P_n) = k$ *for all integers* n *such that*

$$\frac{(k-1)^3+(k-1)^2-2(k-1)+10}{2} \leq n \leq \frac{k^3+k^2+4}{2} \qquad \text{if } k \text{ is odd}$$

$$\frac{(k-1)^3+(k-1)^2+6}{2} \leq n \leq \frac{k^3+k^2-2k+8}{2} \qquad \text{if } k \text{ is even.}$$

14.5   Recognizable Colorings of Trees

In the preceding section, recognizable colorings of paths were considered. Now we consider recognizable colorings of trees more generally. Let T be a tree of order n having n_i vertices of degree i for $i \geq 1$. For each integer $n \geq 2$, let $D(n)$ be the maximum recognition number among all trees of order n and $d(n)$ the minimum recognition number among all trees of order n. That is, if $\mathscr{T}_n$ is the set of all trees of order n, then

$$D(n) = \max \{\mathrm{rn}(T) : T \in \mathscr{T}_n\}$$

$$d(n) = \min \{\mathrm{rn}(T) : T \in \mathscr{T}_n\}.$$

Since no tree of order $n \geq 3$ has recognition number n by Proposition 14.1.4, it follows that $2 \leq d(n) \leq D(n) \leq n-1$ when $n \geq 3$. It is clear that $d(2) = D(2) = 2$. Since it readily follows that the star $K_{1,n-1}$ of order $n \geq 3$ has recognition number $n - 1$, we have the following.

Observation 14.5.1. *For each integer $n \geq 3$, $D(n) = n - 1$.*

It is known that if T is a tree of order n having n_i vertices of degree i for $i \geq 1$, then

$$n_1 = 2 + n_3 + 2n_4 + 3n_5 + 4n_6 + \ldots \qquad (14.2)$$

(see [7, p. 71], for example). By Theorem 14.1.5, if c is a recognizable k-coloring of a connected graph G of order at least 3, then G contains at most k^2 end-vertices and at most $\frac{k^3+k^2}{2}$ vertices of degree 2. It then follows by (14.2) that if T is a tree of order n with $\mathrm{rn}(T) = k$, then

$$n \leq k^2 + \frac{k^3 + k^2}{2} + (k^2 - 2) = \frac{k^3 + 5k^2 - 4}{2}.$$

For example, if T is a tree of order n with $\mathrm{rn}(T) = 2$, then $n \leq 12$. The tree T shown in Fig. 14.7 has order 12 with $\mathrm{rn}(T) = 2$. Observe that T has $2^2 = 4$ end-vertices, $\frac{2^3+2^2}{2} = 6$ vertices of degree 2 and $2^2 - 2 = 2$ vertices of degree 3. In fact, if $2 \leq n \leq 12$, then $d(n) = 2$.

Fig. 14.7 A tree of order 12 with $rn(T) = 2$

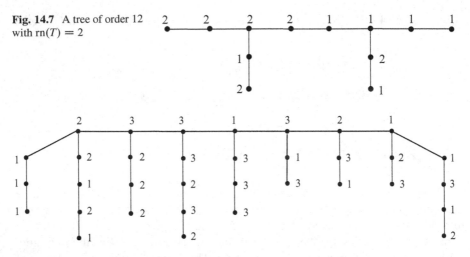

Fig. 14.8 A tree T of order 34 with $rn(T) = 3$

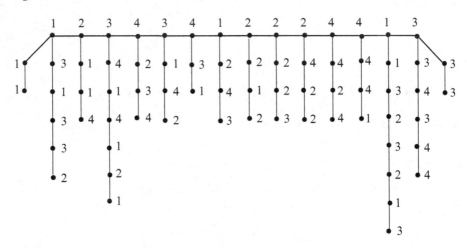

Fig. 14.9 A tree T of order 70 with $rn(T) = 4$

If T is a tree of order n with $rn(T) = 3$, then $n \leq 34$. The tree T of Fig. 14.8 has order 34 and $rn(T) = 3$. This tree T contains 9 vertices of degree 1, 18 vertices of degree 2 and 7 vertices of degree 3. Therefore, $d(34) = 3$.

Also, if T is a tree of order n with $rn(T) = 4$, then $n \leq 70$. The tree T of Fig. 14.9 has order 70 and $rn(T) = 4$. This tree T contains 16 vertices of degree 1, 40 vertices of degree 2 and 14 vertices of degree 3. This shows that $d(70) = 4$.

In general, we have the following conjecture.

Conjecture 14.5.2 ([19]). *For each integer $n \geq 3$, the minimum recognition number among all trees of order n is the unique integer k such that*

$$\frac{(k-1)^3 + 5(k-1)^2 - 2}{2} \leq n \leq \frac{k^3 + 5k^2 - 4}{2}.$$

It is easy to see, however, that the minimum recognition number among all trees of order n is bounded below by the integer k described in Conjecture 14.5.2.

References

1. Aigner, M., Triesch, E., Tuza, Z.: Irregular assignments and vertex-distinguishing edge-colorings of graphs. In: Combinatorics' 90, pp. 1–9. Elsevier Science, New York (1992)
2. Anderson, M., Barrientos, C., Brigham, R.C., Carrington, J.R., Kronman, M., Vitray, R.P., Yellen, J.: Irregular colorings of some graph classes. Bull. Inst. Combin. Appl. **55**, 105–119 (2009)
3. Anderson, M., Vitray, R.P., Yellen, J.: Irregular colorings of regular graphs. Discrete Math. **312**, 2329–2336 (2012)
4. Andrews, E., Lumduanhom, C., Zhang, P.: From combinatorial problems to graph colorings. J. Combin. Math. Combin. Comput. **90**, 75–95 (2014)
5. Bazgan, C., Harkat-Benhamdine, A., Li, H., Woźniak, M.: On the vertex-distinguishing proper edge-colorings of graphs. J. Combin. Theory Ser. B. **75**, 288–301 (1999)
6. Bazgan, C., Harkat-Benhamdine, A., Li, H., Woźniak, M.: A note on the vertex-distinguishing proper coloring of graphs with large minimum degree. Graph theory (Kazimierz Dolny, 1997). Discrete Math. **236**, 37–42 (2001)
7. Benjamin, A., Chartrand, G., Zhang, P.: The Fascinating World of Graph Theory. Princeton University Press, Princeton, NJ (2015)
8. Bi, Z., Byers, A., English, S., Laforge, E., Zhang, P.: Graceful colorings of graphs. J. Combin. Math. Combin. Comput. (to appear)
9. Bi, Z., Byers, A., Zhang, P.: Revisiting graceful labelings of graphs. J. Combin. Math. Combin. Comput. (to appear)
10. Bi, Z., English, S., Hart, I., Zhang, P.: Majestic colorings of graphs. J. Combin. Math. Combin. Comput. (to appear)
11. Brooks, R.L.: On coloring the nodes of a network. Proc. Camb. Philol. Soc. **37**, 194–197 (1941)
12. Burris, A.C., Schelp, R.H.: Vertex-distinguishing proper edge colorings. J. Graph Theory **26**, 73–82 (1997)
13. Černý, J., Horňák, M., Soták, R.: Observability of a graph. Math. Slovaca **46**, 21–31 (1996)
14. Chappell, G.G., Gimbel, J., Hartman, C.: Bounds on the metric and partition dimensions of a graph. Ars Combin. **88**, 349–366 (2008)
15. Chartrand, G., Zhang, P.: Chromatic Graph Theory. Chapman & Hall/CRC Press, Boca Raton (2009)
16. Chartrand, G., Zhang, P.: Discrete Mathematics. Waveland Press, Long Grove, IL (2011)
17. Chartrand, G., Salehi, E., Zhang, P.: On the partition dimension of a graph. Congr. Numer. **131**, 55–66 (1998)
18. Chartrand, G., Salehi, E., Zhang, P.: The partition dimension of a graph. Aequationes Math. **59**, 45–54 (2000)

© The Author 2016
P. Zhang, *A Kaleidoscopic View of Graph Colorings*, SpringerBriefs in Mathematics,
DOI 10.1007/978-3-319-30518-9

19. Chartrand, G., Lesniak, L., VanderJagt, D.W., Zhang, P.: Recognizable colorings of graphs. Discuss. Math. Graph Theory **28**, 35–57 (2008)

20. Chartrand, G., Okamoto, F., Rasmussen, C.W., Zhang, P.: The set chromatic number of a graph. Discuss. Math. Graph Theory **134**, 191–209 (2009)

21. Chartrand, G., Okamoto, F., Salehi, E., Zhang, P.: The multiset chromatic number of a graph. Math. Bohem. **134**, 191–209 (2009)

22. Chartrand, G., Okamoto, F., Zhang, P.: The metric chromatic number of a graph. Australas. J. Combin. **44**, 273–286 (2009)

23. Chartrand, G., Lesniak, L., Zhang, P.: Graphs & Digraphs, 5th edn. Chapman & Hall/CRC, Boca Raton, FL (2010)

24. Chartrand, G., Okamoto, F., Zhang, P.: The sigma chromatic number of a graph. Graphs Combin. **26**, 755–773 (2010)

25. Chartrand, G., Okamoto, F., Zhang, P.: Neighbor-distinguishing vertex colorings of graphs. J. Combin. Math. Combin. Comput. **74**, 223–251 (2010)

26. Chartrand, G., Phinezy, B., Zhang, P.: On closed modular colorings of regular graphs. Bull. Inst. Combin. Appl. **66**, 7–32 (2012)

27. Chartrand, G., English, S., Zhang, P.: Binomial colorings of graphs. Bull. Inst. Combin. Appl. (to appear)

28. Chartrand, G., English, S., Zhang, P.: Kaleidoscopic colorings of graphs. Preprint

29. Chvátal, V.: Some relations among invariants of graphs. Czech. Math. J. **21**, 366–368 (1971)

30. English, S., Zhang, P.: On graceful colorings of trees. Preprint

31. Euler, L.: Solutio problematis ad geometriam situs pertinentis. Comment. Academiae Sci. I. Petropolitanae **8**, 128–140 (1736)

32. Favaron, O., Li, H., Schelp, R.H.: Strong edge colorings of graphs. Discrete Math. **159**, 103–109 (1996)

33. Fertin, G., Raspaud, A., Reed, B.: On star coloring of graphs. In: Graph-Theoretic Concepts in Computer Science. Lecture Notes in Computer Science, vol. 2204, pp. 140-153. Springer, Berlin (2001)

34. Fertin, G., Godard, E., Raspaud, A.: Acyclic and k-distance coloring of the grid. Inform. Process. Lett. **87**, 51–58 (2003)

35. Fertin, G., Raspaud, A., Reed, B.: Star coloring of graphs. J. Graph Theory **47**, 163–182 (2004)

36. Finck, H.J.: On the chromatic numbers of a graph and its complement. In: Theory of Graphs, pp. 99–113. Academic Press, New York (1968)

37. Frank, O., Harary, F., Plantholt, M.: The line-distinguishing chromatic number of a graph. Ars Combin. **14**, 241–252 (1982)

38. Fujie-Okamoto, F., Will, T.G.: Efficient computation of the modular chromatic numbers of trees. J. Combin. Math. Combin. Comput. **82**, 77–86 (2012)

39. Gallian, J.A.: A dynamic survey of graph labeling. Electron. J. Combin. **17**, #DS6 (2014)

40. Gera, R., Okamoto, F., Rasmussen, C.W., Zhang, P.: Set colorings in perfect graphs. Math. Bohem. **136**, 61–68 (2011)

41. Golomb, S.W.: How to number a graph. In: Graph Theory and Computing, pp. 23–37. Academic Press, New York (1972)

42. Graham, R.L., Sloane, N.J.A.: On addition bases and harmonious graphs. SIAM J. Alg. Disc. Math. **1**, 382–404 (1980)

43. Harary, F., Plantholt, M.: Graphs with the line-distinguishing chromatic number equal to the usual one. Utilitas Math. **23**, 201–207 (1983)

44. Harary, F., Plantholt, M.: The point-distinguishing chromatic index. In: Graphs and Applications, pp. 147–162. Wiley, New York (1985)

45. Hopcroft, J.E., Krishnamoorthy, M.S.: On the harmonious coloring of graphs. SIAM J. Algebraic Discrete Methods **4**, 306–311 (1983)

46. Horňák, M., Soták, R.: Observability of complete multipartite graphs with equipotent parts. Ars Combin. **41**, 289–301 (1995)

47. Horňák, M., Soták, R.: Asymptotic behaviour of the observability of Q_n. Discrete Math. **176**, 139–148 (1997)

48. König, D.: Über Graphen und ihre Anwendung auf Determinantentheorie und Mengenlehre. Math. Ann. **77**, 453–465 (1916)
49. Lee, S.M., Mitchem, J.: An upper bound for the harmonious chromatic number of a graph. J. Graph Theory **11**, 565–567 (1987)
50. Lu, Z.: On an upper bound for the harmonious chromatic number of a graph. J. Graph Theory **15**, 345–347 (1991)
51. McDiarmid, C., Luo, X.H.: Upper bounds for harmonious colorings. J. Graph Theory **15**, 629–636 (1991)
52. Nordhaus, E.A., Gaddum, J.W.: On complementary graphs. Am. Math. Mon. **63**, 175–177 (1956)
53. Okamoto, F., Zhang, P.: A note on 2-distance chromatic numbers of graphs. AKCE Int. J. Graphs Combin. **7**, 5–9 (2010)
54. Okamoto, F., Rasmussen, C.W., Zhang, P.: Set vertex colorings and joins of graphs. Czech. Math. J. **59**, 929–941 (2009)
55. Okamoto, F., Salehi, E., Zhang, P.: On modular colorings of caterpillars Congr. Numer. **197**, 213–220 (2009)
56. Okamoto, F., Salehi, E., Zhang, P.: A checkerboard problem and modular colorings of graphs. Bull. Inst. Combin. Appl. **58**, 29–47 (2010)
57. Okamoto, F., Salehi, E., Zhang, P.: A solution to the checkerboard problem. Int. J. Comput. Appl. Math. **5**, 447–458 (2010)
58. Okamoto, F., Salehi, E., Zhang, P.: On multiset colorings of graphs. Discuss. Math. Graph Theory **30**, 137–153 (2010)
59. Petersen, J.: Die Theorie der regulären Graphen. Acta Math. **15**, 193–220 (1891)
60. Phinezy, B.: Variations on a graph coloring theme. Ph.D. Dissertation, Western Michigan University (2012)
61. Phinezy, B., Zhang, P.: On closed modular colorings of trees. Discuss. Math. Graph Theory **33**, 411–428 (2012)
62. Phinezy, B., Zhang, P.: On closed modular colorings of rooted trees. Involve **66**, 7–32 (2012)
63. Phinezy, B., Zhang, P.: From puzzles to graphs to colorings. Congr. Numer. **214**, 103–119 (2012)
64. Radcliffe, M., Zhang, P.: On irregular colorings of graphs. AKCE Int. J. Graphs Comb. **3**, 175–191 (2006)
65. Radcliffe, M., Zhang, P.: Irregular colorings of graphs. Bull. Inst. Combin. Appl. **49**, 41–59 (2007)
66. Radcliffe, M., Okamoto, F., Zhang, P.: On the irregular chromatic number of a graph. Congr. Numer. **181**, 129–150 (2006)
67. Rosa, A.: On certain valuations of the vertices of a graph. In: Theory of Graphs, pp. 349–355. Gordon and Breach, New York (1967)
68. Saenpholphat, V., Zhang, P.: Connected partition dimensions of graphs. Discuss. Math. Graph Theory **22**, 305–323 (2002)
69. Salvi, N.Z.: A note on the line-distinguishing chromatic number and the chromatic index of a graph. J. Graph Theory **17**, 589–591 (1993)
70. Stewart, B.M.: On a theorem of Nordhaus and Gaddum. J. Combin. Theory **6**, 217–218 (1969)
71. Sutner, K.: Linear cellular automata and the Garden-of-Eden. Math. Intell. **11**, 49–53 (1989)
72. Tait, P.G.: Remarks on the colouring of maps Proc. Roy. Soc. Edinb. **10**, 501–503, 729 (1880)
73. Tomescu, I., Javaid, I., Slamin: On the partition dimension and connected partition dimension of wheels. Ars Combin. **84**, 311–317 (2007)
74. Vizing, V.G.: On an estimate of the chromatic class of a p-graph (Russian). Diskret. Analiz. **3**, 25–30 (1964)
75. Wilson, R.: Four Colors Suffice: How the Map Problem Was Solved. Princeton University Press, Princeton, NJ (2002)
76. Zhang, P.: Color-Induced Graph Colorings. Springer, New York (2015)
77. Zhang, Z., Liu, L., Wang, J.: Adjacent strong edge coloring of graphs. Appl. Math. Lett. **d15**, 623–626 (2002)

Index

© The Author 2016
P. Zhang, *A Kaleidoscopic View of Graph Colorings*, SpringerBriefs in Mathematics,
DOI 10.1007/978-3-319-30518-9

Printed in the United States
By Bookmasters